Mercury
ultra-trace analysis

Hydrargyrum

From "diphenylthiocarbozone"
to "atomic absorption."

3rd edition

3rd edition

Dr. Paul H. Ramses Gouda
C.Chem., P.R.M.D., Ph.D.

Mercury analysis

A look at the analytical laboratory techniques; with emphasis on atomic absorption.

**A chemist manual,
a laboratory reference,
and a university textbook.**

www.goudabooks.com

www.optimumgreen.com

Front cover by Pelé Gouda

Video animation, cartoon illustration and graphic design.
pelegouda@gmail.com

Copyright © 2013 by Paul Gouda.

All rights reserved. No part of this book may be used, reproduced or transmitted in any form or by any means, electronic, graphic or mechanical, including photocopying, recording, or by any information storage retrieval system, without a written permission from the author, except in the case of brief quotations embodied in academic or critical articles and reviews.

ISBN: 978-1-4759-9462-9 (sc)
ISBN: 978-1-4759-9463-6 (ebk)

Printed in the United States of America

iUniverse rev. date: 06/18/2013

Copyright registration USA & Canada 2013

Canadian Intellectual Property Office.
Certificate of registration of copyright number: **1104260**
Paul Gouda, Owner & Author.
Issued pursuant to sections 49 & 53 of the Copyright Act.

Library and Archives Canada Cataloguing in Publication
Gouda, Paul
Hydrargyrum : ultra trace analysis / Paul Gouda.
1. Mercury--Analysis. 2. Ultratrace analysis. I. Title.

QD181.H6G68 2011 546'.6636 C2011-901195-6

American Library of Congress Control Number: **2013910813**

Dr. Paul Gouda, at his laboratory office at EPA during a media release report.

"A valuable reference for every laboratory, university, and science library. The author, who is an authority on the subject, and who has created several methods that were named after him, has provided an in-depth analysis of mercury ultra trace analysis."

> Dr. Nadi Habib, Ed.D.
> Science News Gazette, 2008

Because of the dynamic nature of the internet, any web addresses or links contained in this book may have changed since publication and may no longer be valid.

                                                    The publisher

This book was first produced by the author as a bench manual and a chemist reference back in the mid 1990s, while serving as an analytical chemist at BRM, OGC and EPL.

In the late 1990s, the author posted most of the material of this book in the form of several chapters and papers online, via several internet science sites, and the feedback he received encouraged producing a complete reference on this topic. In 1994, it witnessed the production of its first complete and printed manual.

Most of my material, both the online papers and the printed manual and articles, were more than once summarized and at times was basically "cut & pasted" verbatim, and reintroduced. We know only of one case in the 1990s when credit or recognition of Dr. Gouda's research and experimentations produced via this book, was officially acknowledged. The author, in an interview with OGCC science gazette in 2001 stated: "in all honesty, that didn't bother me at all. After all, the material served an educational and a positive purpose and that's what counts. And, I encourage the use of this book's material."

This material was then published by Great Canadian Books in 2003 as a university textbook and a chemist manual. It was a special, limited edition intended to meet the semester demands of that year. The book was not as large or as detailed as this edition, and it was priced by the publisher at $95 and was sold out in 6 months. This 3rd edition is made extremely affordable for a wide distribution.

Encouraged by the scholastic interest expressed by several universities and analytical laboratories, this 3rd edition was published by Penguin Group Books, via its branch iuniverse, 2013, to offer students and public libraries an affordable copy.

The book offers a complete coverage of the subject.

iUniverse books may be ordered through booksellers or by contacting: iUniverse
1663 Liberty Drive.
Bloomington, IN 47403
1-800-288-4677
International: 001-812-330-2909

You can also order this book at your local branch of any of the following bookstores:

  - **Chapters, Coles, Indigo; Canada.**

  - **Barnes & Noble; USA.**

In the absence of shelf-copies at your local bookstore, you may still order this book at any "Barnes & Noble" or any "Chapters – Coles – Indigo" bookstore, via their in-store database search & order kiosks.

This book is also available online at:

    www.amazon.com
    www.bn.com   "**Barnes & Noble**"
    www.chapters.ca
        "**Chapters - Coles - Indigo**"
    www.biblio.com   (Harvard bookseller)
    www.iuniverse.com
    www.abe.com
    www.alibris.com
    www.biggerbooks.com
    www.bookdepository.com
    www.goodreads.com
    www.buy.com - and several other online bookstores.

Table of contents

Preface
Synopsis
Chemical properties
Biomethylation of mercury
Electrode potentials

Calciunation method
Gravimetric bench-chemistry methods
Titrimetric methods
Volumetric methods
Dithizone method

Diphenylthiocarbazone, a closer look
Neutron activation method
Thiocynate methode
Mercury amalgamation trap

Inorganic forms
Specific approached to organic samples

Cold Vapour atomic absorption, theory & principal
Cationic & anionic interferences
Flame stoichiometry

Cold vapour atomic fluorescence spectroscopy
Thermal decomposition

Methylmercury
Phenyl mercuric acetate and methylmercuric chloride
The dithizone extraction / gas-liquid chromatography
Electron capture detection

Cold vapour atomic absorption issues
Interference by certain volatile organic materials

Chlorides and oxides of nitrogen
Diatomic gases

Phenylmercuric acetatate,
chloro-methoxypropylmercuric acetate
and alkyl mercuric compounds

Correlation coefficient
AAS setting & the calibration curve
Mercury salts

Phenylmercuric acetatate
Chloro-methoxypropylmercuric acetate
Alkyl mercuric compounds
Volatile organic material that absorb radiation

AAS setting
Quality control
Possible complications
Conclusion

About the author
Other books by the author

The author, Dr. Paul H. Ramses Gouda, holds two doctorates, with research papers and books in pharmaceutical neurochemistry, ultra trace analysis of food, environmental and drug samples, and hormonal compounds manipulation.

Preface

The presence of mercury - in several forms, and in any substance that we deal with, is a concern raised by the possible toxicity of many of its forms. Environmental analytical laboratories regularly analyze soil, lake water, potable water, seafood, pharmaceutical products, and several other types of samples for mercury contamination.

Such contamination is usually caused by industrial and commercial pollution. As a result, I have personally seen alarming levels of toxic mercury in fish and water samples I analyzed during the past two decades.

This book will deal with the technical aspect of laboratory analysis SOPs and will examine several options.

The emphasis here will be on cold vapour atomic absorption technique whereby mercury ions in an acidic solution are reduced by reaction with stannous chloride to ground state atoms.

The solution is vigorously stirred until the mercury vapor over the solution reaches equilibrium with the mercury left in the solution.

The total Mercury atoms (atomic mercury vapor) is driven {by use of purge gas: nitrogen or argon} into a quartz absorption cell located in the optical path of AAS.

Hg, being in a ground atomic state is amenable to atomic absorption of radiation and Hg atoms are monitored at 253.7 nm wavelength.
This is the basis theory of the SOP. We'll get into details later.

This book will deal specifically with the laboratory treatment, digestions, complications, side effect and undesired by-products "compounds" that the bench chemist may face and must be aware of, and the tips and advices drawn from personal analytical laboratory experience.

A major part of the paper will reflect on a solid atomic absorption approach. I will still have to touch on related issues that constitute a significant role in the overall success of the analytical chemist in understanding what is happening in the sample at all times.

I also want to comment that while mercury use, as well as many of its treatment techniques are in fact as ancient as the Pharaohs and the Romans; and while recorded history proves that the Egyptians followed by the Greeks and Romans used mercury for cosmetic and medical preparations over five thousand years ago; we are now talking about ppt level, which mocks the expression "splitting hair."

Indeed, the basic structure of a wet procedure still being used today involving mercury and gold amalgamation is recorded in hieroglyphic and is dated back to year 2800 B.C. It was also the SOP utilized in 1600 BC when Aristotle referred to mercury as liquid silver. This is the root-concept of the method that was recently developed to ppt detection limit with contemporary instrumentation.

This book will lead you through the bench – wet chemistry steps prior to and during the instrumentation component of the SOPs, and will also go through other wet SOPs we still need to use, to understand and to manipulate. The wet – or bench part of the analysis is most significant. Instrumentation can only read what the chemist has written in that test tube.

Synopsis

Mercury minerals in ore deposits in nature consist mainly of corderiote ($Hg_3S_2CL_2$), livingstonite ($HgSb_4S_7$), montroydite (HgO), terlinguaite (Hg_2OCL), cinnabar HgS, calomel Hg_2CL_2, coloradoite HgTe, and teimannite HgSe.

Mercury exists in two oxidation states: mercurous (valence +1) and mercuric (valence +2). Organic compounds of weak reducing activity such as amines, aldehydes and ketones often break mercury compounds to compounds of lower oxidation state and metal.

Chemical properties

Mercury does not react with most acids, such as dilute sulfuric, although oxidizing acids such as concentrated sulfuric acid and nitric acid or aqua regia dissolve it to give sulfate, nitrate and chloride salts. Like silver, mercury reacts with atmospheric hydrogen sulfide.

Mercury dissolves many other metals such as gold and silver to form amalgams. Iron is an exception and iron flasks have been traditionally used to transfer mercury. Several other first row transition metals with the exception of manganese, copper and zinc are reluctant to form amalgams. Other elements that do not readily form amalgams with mercury include platinum and a few other metals. Sodium amalgam is a common reducing agent in organic synthesis, and is used in high-pressure sodium lamps.

Mercury readily combines with alluminium upon contact, to form an mamagam that destroys the aluminium oxide layer which protects metallic aluminium from oxidizing in-depth.

In the mercurous form, it is precipitated by hydrochloric acid as white mercurous chloride Hg_2CL_2. This compound is changed by ammonium chloride to the black precipitate of metallic mercury and ammono-mercuric chloride.

Determination of Hg

Determination of mercury went through several stages of development. However, depending on the sample, the desired DL and the expected Hg level, some of the older methods may constitute a suitable approach.

Determination of mercury by electrolysis:

150 ml of neutral or slightly acidic solution of mercuric of mercurous salt is used.

Titrimetric methods:

0.5 g of ground ore is placed in an Erlenmeyer flask + 5 ml hydrochloric acid. apply heat to 40C for 10 minutes.

Add 3 ml nitric acid and continue to digest for about 10 minutes. If lead is present, add 5 ml of conc. Sulfuric acid. dilute with 15 ml water, ten add ammonia until the liquid is slightly alkaline. Acidify faintly with nitric acid, filter, receiving the filtrate into a beaker. Wash.

Add 1 ml nitric acid that has been exposed to light enough to turn brownish. Titrate with standard solution of potassium iodide until a drop of the liquid brought into contact with a drop of starch liquor, on a hot plate, shows a faint bluish tinge.

Titrate to end point and deduct 0.5 ml from the burette reading, and multiply he remaining milliliter used by the percentage value of 1 ml in mercury to obtain the percentage in the ore. Standard potassium iodide solution contains 8.3 g of the salt per liter. Standardize against pure mercuric chloride. Dissolve a weighed salt in water, add 2 ml conc.discoloured nitric acid and titrate as indicated.

1 ml of standard solution = 0.005 g of mercury.

Notes:

Titrimetric methods, notes

Mercurous ion is oxidized by potassium permanganate or by ceric sulfate in excess.

Potassium iodate is used for direct oxidation of calomel under iodine monochloride conditions.

Mercury can be precipitated as zinc mercuric thiocyanate; and the thiocyanate is then titrated with standard iodate solution.

Mercurous mercury can be titrated with standard chloride or bromide solution using bromphenol blue as adsorption indicator.

Mercuric chloride can be titrated with standard potassium iodide to the appearance of a permanent red turbidity.

Mercuric ion can be precipitated as mercuric pyridine bichromate and weighed, or the bichromate can be titrated by conventional methods.

Thiocyanate methods

0.1N thiocynate solution is prepared: 7.4 g of NH4CNS or 9.2 g of KCNS in water, diluted to 1 L.

You can confirm this by standardizing it against a standard silver solution containing 0.01079 g silver / L.

Measure 40 ml of the silver solution into a beaker or Erlenmeyer flask, and dilute to 100 ml. Add ferric indicator and titrate with thiocyanate solution. A temporary red colour will temporarily appear with each addition. The end point is the drop of excess thiocynate that produces a permanent faint red colour.

The thiousulfate can be standardized against pure mercury dissolved in dilute HNO_3 using the same procedure.

A sample containing 0.1 to 0.5 g of mercury placed in a flask is decomposed by adding 10 ml of dilute H_2SO_4 1:1 and 0.5 g of $KMnO_4$ crystals. The mixture is agitated and heated to fumes.

The solution is cooled and diluted to 50 ml with cold water, then boiled and the MnO_2 dissolved by slowly adding a few crystals of oxalic acid.

Filter and wash recovered residue with dilute 1:10 H_2SO_4. The sulfide group is now precipitated with H_2S and filtered off. Transfer the precipitate to a flask with a short-stemmed funnel and digest it with dilute 2:1 HNO_3 first, then with hot water, and filter the HgS washed with dilute 1:1 HNO_3.

The HgS is transferred to a flask with 5 ml of hot water and then 5 ml of strong H_2SO_4 and 0.5 g of $KMnO_4$ and heat the mixture to fumes. Add oxalic acid crystals until the MnO_2 dissolves, and the mixture is heated again to fumes to expel excess oxalic acid.

The solution is cooled, diluted to 100 ml. It should now be clear.

Add 5 ml of saturated solution of ferric ammonium sulfate that has been acidified with HNO_3.

Titrate with 0.1 N thiocyanate solution. Calculate your analysis result based on:

1 ml 0.1 N thiocyanate = 0.01003 g mercury

Colorimetric methods

Dithizone:

The concept here is the formation of a yellowish orange colour with dithizone by mercury II compounds in the pH range 1-2.

The complex formed is insoluble in water but is soluble in CCL_4 and $CHCL_3$. Use sulfuric acid in acidification because Chloride in moderate concentration interferes. This approach also requires the absence of organic matter from the solution to avoid interference.

To prepare the dithizone solution, dissolve in I L of analytical reagent chloroform, 10 mg of specially-purified dithizone.

A commercial grade can be purified this way:

- dissolve 1 g of dithizone in 60 ml of chloroform, filtering if needed.

- Extract with 4 X 100 ml portions of 1:100 redistilled concentrated ammonia.

- Discard the chloroform layer.

- Filter combined aqueous extracts through cotton into a large separatory funnel.

- Acidify to litmus with 1:1 HCL. The dithizone precipitate.

- Extract it with 3 X 20 ml portions of chloroform. Combine the extracts.

- Wash them with water and discard the water.

- Evaporate the chloroform solution on a steam bath. Dry the purified dithizone for 1-2 hours at 50 C in vacuo.

- sulfuric acid solution should be 1:17 (to 170 ml water in a flask, add 10 ml concentrated H_2SO_4 slowly, while stirring.)

Note that this procedure is influenced by sensitivity of colour to light. Bench work needs to consider this fact by dimming the light or using low and artificial light.

- place in a separator funnel in aliquot of sample containing 0.01 to 0.02 mg of mercury.

- neutralize with ammonia.

- dilute to 20 ml and add 20 ml of 1:17 H_2SO_4

- add 5 ml of dithizone solution in $CHCL_3$

- shake for a minute and let the solvent separate.

- if the colour of the solution is not green, indicating excess reagent (or red-violet, indicating the presence of copper) add another 5 ml of dithizone solution and shake again. Repeat this addition if needed until the green colour "A" or the red-violet colour "B" appears.

A - if copper is absent, based on the colour as indicated above: draw off the solvent layer. If it is not clear, centrifuge it in a stoppered tube. Minimize exposure to light and read the transmittance at once. Read the mercury dithizonate at 500 mu and check by reading the excess dithizone at 625 mu

B- copper is present: draw off the chloroform layer (discarding the aqueous layer) and shake it with 10 ml 6 N H_2SO_4. Adjust pH to 1.5 – 2 by slow addition of 1:1 ammonia, then add 5 ml of .001 M sodium ethylenediaminetetraacetate solution. Shake again, and discard the chloroform layer.

- add 20 ml of 1:17 H_2SO_4 to the aqueous layer and 5 ml of dithizone solution in chloroform. Proceed from here according to the procedure.

Lab notes:

Isotopes

There are seven stable isotopes of mercury with ^{202}Hg being the most abundant (29.86%). The longest-lived radioistopes are ^{194}Hg with a half-life of 444 years, and ^{203}Hg with a half-life of 46.612 days. Most of the remaining radioisotopes have half-lives that are less than a day.

The toxicity of mercury compounds is not a recent discovery either! Galen, who died about 200 A.D. wrote about the toxicity of mercurials. As to the therapeutic uses of mercury compounds, the "Chemical technology reference" makes this statement: "... in the 13th century, as a result of Arabian influence, such therapeutic uses of mercury were finally recognized by Western Europe."

Some of these uses are specifically supported by recorded history in Egypt 5000 years ago, and in China and India dating over 4000 years ago.

Mercuric sulphide has been historically known to decompose by heat, in contact with air in different forms of furnaces, to provide condensation of the vapour, clearly at a temperature above the level mercury volatilization. The essential reaction is:

$$HgS + O_2 = Hg + SO_2$$

However, the liberation of mercury in the absence of air, by the intervention of lime or iron provides this reaction:

$$4HgS + 4CaO = 4Hg + 3CaS + CaSO_4$$

$$HgS + Fe = Hg + FeS$$

Mercury analysis and treatment has surely witnessed a sharp peak of advancement since these days.

The wide attention and the pressure to achieve higher detection limit was driven mainly by the focus on the toxicity of mercury, especially with incidents such as the famous industrial Japanese tragedy.

Having said that, it seems however - at least at times – that the media attention surrounding the toxicity of mercury is associated with more than one misconception.

Contrary to common belief, metallic mercury is not highly toxic. In fact, a suicidal attempt by injection of two grams of mercury failed to produce more than a rather minor trouble.

Meanwhile, organic mercury compounds [and fumes, such as dimethyl-mercury $(CH_3)Hg$] are extremely toxic. Such compounds have been found in fish. PMA [phenyl mercury acetate] $C_6H_5HgCOOCH_3$ is also very poisonous.

Modern technology has also utilized mercury in a negative way with the use of mercury fluminate $Hg(ONC)_2$ as a detonator for explosives.

The discovery of biomethylation of mercury and the realization of the role of certain chemical moieties not only as toxic sources but also as potential carcinogens and mutagens are behind the rapid development of powerful analytical instrumentation capable of detection of mercury compounds in ranges of fractional parts per billion {ppb} – and in fact, new instrumentation can detect Hg at the ppt level.

Within a rather short period of time, techniques, including chromatography, micrometry, radiometry, spectrography and titrimetry, were developed considerably.

The liquid silver in action:

Mercury, in the +2 state is frequently present as the simple ion Hg^{2+}. It is however usually found in the form of complex ions, insoluble solids or weak salts. In fact, in a solution of the weak salt mercuric chloride, the concentration of Hg^{2+} is much smaller than the concentration of non-dissociated $HgCL_2$ molecules. With excess chloride ion, the complex $HgCL_3^-$ and $HgCL_4^{2-}$ are also formed.

In Ammonia solutions complex ions containing one to four NH_3 molecules are expected.

For a complete dissociation of $Hg(NH_3)_4^{2+}$ the constant is 5.2×10^{-20}. The complex $Hg(CN)_4^{-2}$ is even more stable.

Mercuric sulphide as found in nature is a red salt, however, when it is passed through a mercuric solution, a black precipitate of HgS is obtained. The colour difference is due to difference in crystal structure. The solubility product of black HgS is very low $\{1.6 \times 10^{-54}\}$. Such insoluble sulphide will not dissolve even in boiling nitric acid. Aqua regia however, which supplies both nitrate for oxidizing the sulphide and chloride for complexing the mercuric, does take it into solution.

The electrode potentials are so close that any reducing agent is able to reduce mercuric ion to mercurous ion can also reduce mercurous ion to mercury. Thus, if a limited amount of reducing agent solution such as Sn^{2+} {stannous ion} is added to a mercuric solution, only Hg_2^{2+} is formed, but if Sn^{2+} is added in excess, the reduction goes all the way to Hg.

$2 Hg^{2+} + 2 e^- = Hg_2^{2+}$

$Hg_2^{2+} + 2 e^- = 2 Hg$

A look at mercury salts

mercuric sulphide or mercury II sulphide (HgS), mercuric oxide or mercury II oxide (HgO) and mercuric chloride or mercury II chloride (HgCl$_2$).

Mercury salts are usually white powders or crystals. Mercuric sulphide, which is normally red, is an exception.

Salts:

(CH$_3$COO)$_2$Hg

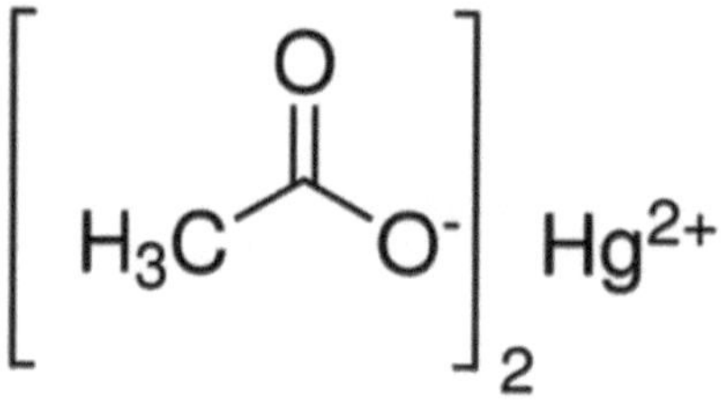

Mercury(II) acetate

Mercury(II) benzoate hydrate

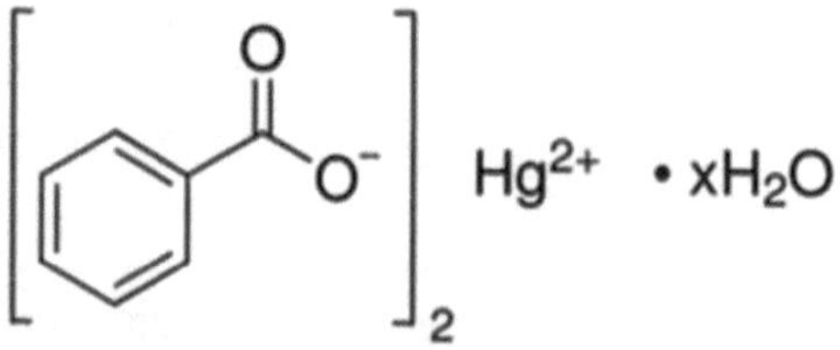

Hg(OCOC$_6$H$_5$)$_2$ · xH$_2$O

There are several other mercury salts, some of which are significant as far the bench chemist role is concerned:

Mercury(II) bromide
HgBr$_2$

Mercury(II) bromide
$HgBr_2$

Mercury(I) chloride
Hg_2Cl_2

Mercury(II) chloride
$HgCl_2$

Mercury(II) cyanide
$Hg(CN)_2$

Mercury(II) fluoride
HgF_2

Mercury(II) iodate
$Hg(IO_3)_2$

Mercury(I) iodide
Hg_2I_2

Mercury(II) iodide
HgI_2

Mercury(II) perchlorate hydrate
$Hg(ClO_4)_2 \cdot xH_2O$

Mercury(I) sulfate
Hg_2SO_4

Mercury(II) tetrathiocyanatocobaltate(II)
$HgCo(SCN)_4$

Mercury(II) thiocyanate
$Hg(SCN)_2$

Note: iorganic murcury:

Inorganic mercury compounds are formed when mercury combines with elements other than carbon, such as chlorine, sulfur, or oxygen. Elemental mercury is a form of inorganic mercury.

Mercury binds in two states:

Hg^+ (mercurous or mercury I), and
Hg^{2+} (mercuric or mercury II).

Organic mercury is found in three compounds: methyl-, ethyl- and phenylmercury. Methylmercury is the most toxic, formed by microorganisms from elemental mercury found in the environment via human or natural sources.

Mercury analysis, now and then

As I indicated earlier, mercury analysis was developed considerably within the past few years to meet the new and much tighter environmental specifications. Not too long ago, I reported Hg in ppm using conventional FAA "flame atomic absorption". Mercury analysis by flame absorption using a concentration of 5 mg l^{-1} in solution produced an absorbency signal of only .0050. This is obviously is not good enough for today's specification level.

In fact, not too long ago I analyzed several sediment and industrial sludge samples completely in a fume-hood in the wet lab, with just a few test-tubes, a hot plate, beakers, flasks and funnels; applying titrimetric / volumetric approaches - reporting a final concentration in g/l & %. Now a days such concentration is rare.

Gravimtetric procedures:

Hard to believe that only a few years ago colorimetric tests such as cinnabar / black HgS test, and $HgNH_2CL$ tests were commonly accepted as orthodox methodology!

Titration / volumetric as well as gravimetric methods to analyze Hg were justified by the high level of mercury present in alloys and industrial samples. In **the calciunation method**, mercury and a deficiency of hot, concentrated nitric acid react to form mercurous nitrate:

$$6\ Hg + 8\ HNO_3 \rightarrow 3\ Hg_2(NO_3)_2 + 2(NO3)2 + 2NO + 4H_2O$$

Water and nitrogen oxide are driven off
Continued heating drives off NO_2 or generally NOx:

$$Hg(NO_3)_2 \rightarrow 2HgO + 2NO_2$$

In **the hot precipitation method**, sodium carbonate solution is added slowly to a refluxing solution of mercuric chloride, followed by an additional reflux for 1-2 hours. The washed precipitate is then dried.

Another approach {*the EGM method*} is to separate mercurous ion from cations by adding HCL to precipitate white, insoluble $_2H_2g_2CL$. When NH_3 is added to a mixture of this chloride, a black colour appears indicating the formation of Hg and **$HgNH_2CL$**.

If H_2S is added to an acidic solution containing Hg^{2+} {and other elements: e.g. Cd^{2+}, Cu^{2+}, Zn^{2+} ...} insoluble sulfides are precipitated:

black HgS,

yellow CdS,

black CuS.. etc.

"The bench chemist needs to follow up with further test-tube approaches to determine what he has. I'll illustrate."

The residual solution is made basic with NH_3 and white ZnS is formed.

The next step involves a confirmatory test to separate other elements and to isolate mercury. ZnS is dissolved in HCL. Evaporated to dryness and re-precipitated by addition of H_2S in a SO_4^{2-} - $H_2SO_4^-$ buffer.

The separation of HgS, CdS, CuS makes use of the fact that CdS and CuS are soluble in boiling HNO_3 where HgS is not.

Residual HgS can then be confirmed by dissolving it in aqua regia and reducing it with $SnCL_2$ into **Hg_2CL_2 + Hg.**

Hg is then separated as indicated and re-precipitated while Hg in Hg_2Cl_2 is calculated. The total Hg recovery is then achieved.

Such gravimetric bench-chemistry methods for Hg analysis are almost a thing of the past when it comes to environmental samples. However, I must admit, I still now and then resort to such methods as a confirmatory test on high concentration alloy samples and Hg salts such as $HgCl_2$ as dictated by specific SOP standards. It still feels good when a gravimetric straight test-tube recovery of near 100% is obtained!

Volumetric approaches

The titration approach, obviously for high level of concentration, has more than one common execution method.

The tartaric acid & sodium bicarbonate titration method widely adopted by many laboratories until a few years ago, again became unsuitable for today's expectation of DL "Detection Limit." Like most analytical chemists, I personally experimented with more than one method.

The common method involves heating the sample to volatilize mercury and collect it as metal, which is then dissolved in hot HNO_3.

$KMNO_4$ is added to oxidize the mercury, and peroxide to destroy excess permanganate.

Ferric sulphate is added [nitrate indicator] and the solution is titrated with standard potassium thiocynate solution to a faint pink end-point.

Lab notes

Diphenylthiocarbazone ...

The dithizone method

The next stage of development witnessed approaches such as the **diphenylthiocarbazone** [dithizone] method which became commonly utilized. More than one approach was introduced.

A finely powdered sample is treated with sulphuric acid, hydrobromic acid and bromine to give a solution adjusted to pH 4. The solution is treated with dithizone in n-hexane to form mercuric dithizonate. Typical detection limit was reported to be 0.02 ppm.

The ECS diphenylthiocarbazone determination is simply based on mercury ions' reaction with dithizone solution in chloroform producing an orange colour. The development of much more sensitive spectrophotometry contributed to the wider use of the dithizone method; however, certain factors remained to be major obstacles; namely:

1- *Interference* by Pd, Cu, Au, Ag & divalent Pt which react with dithizone in acidic solutions.

2- **Cu in dithizone extract remains in the organic phase while Hg dissolves in the aqueous phase**. Since mercury dithizonate is very photosenitive, it must be measured quickly.

3- The expected *detection limit* even with the aid of a good spectrophotometer providing a light path of 1 cm or longer at 492 nm is still much higher than the concentration level needed to be monitored in order to meet the environmental expectation.

The dithizone method is, in a way, similar in several aspects to the now widely applied cold vapour procedure adopted by (but not invented by) the North American Environmental Agency [EPA] and documented as method # 245.1 & 245.2

The reagents utilized by *the Dithizone method* are potassium permanganate / potassium persulfate, potassium bromide, hydroxylamine hydrochloride, phosphate-carbonate, sodium sulphate, dithizone in chloroform solution and sulphuric acid as indicated below:

5% $KMnO_4$,

5% $K_2S_2O_8$,

40% KBr,

50% $NH_2OH.HCL$,

15% $Na_2HPO_4.12H_2O$ &

3.8% anhydrous K_2CO_3 {extract with 10 ml portions of dithizone until the last portion remains blue, then rinse with $CHCL_3$ to remove excess dithizone},

Na_2SO_4 anhydrous,

60 ml of stock dithizone solution in $CHCL_3$ to produce 1 ml = 6 ug dithizone, H_2SO_4 conc.& 0.25N.

The method is obviously quite lengthy and time consuming; however it has provided the most successful determination of low level mercury specially with organic mercury compounds such as methyl mercuric chloride at concentration level of 250 ug / l.

At this level, a recovery >95% was achieved. At a time when concentration < 1 ppm was considered and reported as BDL "Below Detection Limit," the dithizone set-up was certainly the epitome of ultra-trace analysis.

The neutron activation method was never widely adopted by commercial analytical labs for obvious reasons. The method is based on activating the sample by neutron bombardment to emit gammarays. This still left the analytical industry with a need for a more analytically and economically sound approach.

ppt recovery

The next development of cold vapour / atomic absorption technique was able to secure accurate determination of mercury at trace ppb level. The commonly expected environmental detection limit improved to about 1 ppb.
The EPA 245 cold vapor method secures a detection limit of 0.2 ppb or even lower. This is based on my own analysis record when I documented the initial data using the instrumentations available in the 1990s.

This AAS detection limit can be improved further to as low as 0.01 ppb (ppt level) under certain analytical conditions and using the newly proposed signal enhancement technique.

The approach is based on the fact that mercury forms **an amalgam with Au**.

A collecting surface is inserted into mercury vapor flow then heated to drive off trapped mercury. The procedure is known as MAT *"mercury amalgamation trap"* and uses a thin layer of gold plated on a large surface.

The reagents used are the same as per cold vapor EPA 245, in fact, the procedure utilized the same vapor generator unit introduced by Varian for cold vapor AAS determination with the addition of a tube inside which the collecting sheet is placed.

The procedure is referred to herein for documentary and reference purposes only. It does however, in my opinion, have its share of difficulties including the need for a very long period of collecting time per each determination to deal with the ubiquitous nature of Hg; not to mention that ultra-pure reagents and even a distillation process become a must.

This is only logical since a 99.999 % pure reagent could mean an impurity level of 10 ppm.

Having said that, let's then scan through the cold vapor technique that apparently seems to be our best bet.

The theory behind the cold vapour technique

Mercury ions in an acidic solution are reduced by reaction with stannous chloride to ground state atoms. The solution is vigorously stirred until the mercury vapor over the solution reaches equilibrium with the mercury left in the solution.
The total Mercury atoms (atomic mercury vapor) is driven {by use of purge gas: nitrogen or argon} into a quartz absorption cell located in the optical path of AAS. Hg, being in a ground atomic state is amenable to atomic absorption of radiation and Hg atoms are monitored at 253.7 nm wavelength.

As indicated below, special effort {**during wet preparation as well as during instrumentation**} is essential to achieve good analysis. The educated instrumentation approach must be aimed at maximizing the utilization of the AAS to its fullest: e.g. alteration of spectral bandwidth / gain / cathode lamp current ...etc. - based on sample needs when necessary.

As well, **the digestion scheme** may also have to be altered from one sample to another.

Generally, the adopted wet part of the procedure and a good computerized AA programme constitute a solid guideline. However, having said that, it must be stated that as in any analytical performance, understanding the nature and the chemistry of the sample and altering both the wet and the instrumental approaches taken to treat and to monitor the sample are not only expected but are also imperative.

No flame stoichiometry

The nature of the cold vapour procedure eliminates several common AA concerns associated with flame stoichiometry - including several *cationic & anionic interference* and other common problems that may rise when monitoring the excited atoms in the flame.

At the same time, traditional Hg analysis {as **Hg I** or **Hg II**: $Hg_2^{2+} \longrightarrow Hg^2 + Hg^0$} by flame atomization does not produce the required detection limit sensitive enough for environmental research {typical FAA -Hg recovery is in ppm: > 1 mg/L } and introduces additional C_2H_2 negative side effects.

The cold vapour – recovery, despite the fact that it is a straightforward procedure, yet it does have its "imperfections." In this report, I shall provide examples of practical experience to elucidate the intricacies of the analytical approach and to illustrate the necessity of understanding all active factors involved -e.g. *background interference, chemical and spectral* suppressants or false positives and all other possible analytical scenarios.
This understanding is a must in order to be able to manipulate the conventional and the instrumental analytical approach and thus monitor and capture the right peak.

Today's analytical chemist faces a challenge!

Indeed as with all AAS analytes, one cannot depend on a blind routine application of a parameter method. Hg - CVA is certainly no exception. The variety of samples representing different natures and different active backgrounds necessitates, at times, individual attention given to a specific sample. Without such individual treatment one may only reach a semi-reasonable or in fact a totally wrong figure, and certainly one will not capture the perfect result. Those who understand the nature of instrumental analysis would surely agree that this statement is indeed not a hallow theory, it's a practical fact that must be observed.

A blind routine application of the method, regardless of the sample's matrix, concentration, nature, interfering factors, chemical side-effects and other details of the analysis picture, could mean a meaningless figure. Such figure will vary - for obvious reasons - from an organic to an inorganic sample, vitreous / amorphous solids to crystalline solids, a soil sample to an industrial solution sample, or from a metal to a fish sample.

The specific reagents often used to produce a good result, can cause complications if applied blindly to another sample. For example, sulphuric acid can introduce side reactions that, in turn, would create other reactions ..

$H_2SO_4 + NaCL \rightarrow Na_2SO_4 + HCL$.

Also, the role of $KNO_3 + NaCL$ ... depending on other agents taking direct or indirect active participation, can introduce several cases of unwanted situations....and, the ionization of $KNO_3 = NO_{3-} + K^+$,

Or: $NaCL = Na^+ + CL-$

Or:

$HCL = CL- + H^+$ ..

Or the decomposition of $HgO = Hg + O_2$

and several other possible interference by the dichromate $K_2Cr_2O_7$ or by the ammonium salt $(NH_4)_2Cr_2O_7$ ... all such scenarios must be examined closely from one sample to another.

I will specify a few selected examples of cases I ran into personally and cases that have been reported by other analytical chemists through established experimentation.

These notes are supported by the basic principals of chemistry, documented methodology, established research literature and personal experience. I shall focus in this part of the "paper" on cold vapor approach as being - as stated earlier - the most suitable candidate for this field of analyses. But, first, let's establish a couple of basic facts.

General notes

The most utilized analytical techniques for measuring mercury are cold vapor atomic absorption spectroscopy (CVAA), cold vapor atomic fluorescence spectroscopy (CVAF), and direct analysis by thermal decomposition. Each technique has advantages and disadvantages. This chapter will address this issue.

CVAA

CVAAA is still the most commonly used technique for measuring mercury. It offers a very low detection limit, in the single-digit parts-per-trillion (ppt) range, and several analytical methods that allow for the measurement of mercury in almost any sample matrix.

CVAF

CVAF mercury analyzers include sub-part-per-trillion detection limits and a much wider dynamic range than achieved by CVAA; typically 5 orders of magnitude for CVAF versus 2-3 for CVAA. CVAF instruments are available in two configurations; one employing simple atomic fluorescence and one that employs gold amalgamation to pre-concentrate Hg prior to measurement by atomic fluorescence. The detection limit via the simple fluorescence approach is about 0.1 ppt whereas using the pre-concentration with fluorescence detection can be as low as 0.01 ppt.

Direct Analysis by Thermal Decomposition

This approach eliminates the sample digestion step, offering fast analysis times, and a detection limit of about 0.005 ng. Eliminating digestions means solid samples can typically be run in their native form. For labs that analyze large numbers of solid samples, or that would rather not perform the digestion associated with CVAA and CVAF, direct analysis may be ideal.

Before you can decide which technique is right for you, you need answer a few questions:

Do you need to comply with a regulatory method?

For some laboratories, the decision will be driven solely by the need to comply with a specific regulatory method. e.g. if your lab is required to analyze samples using EPA method 245.1, then you will need to resort to CVAA.

If your lab is not constrained by a regulatory method, the driving force more than likely be such things as:

- The characteristics of your sample matrix (for example, is it a sludge, alloy, food, or a liquid etc.)
- The detection limit you need to reach in that matrix.
- Your lab's – including the staff's capabilities regarding sample digestion.
- Your budgetary constraints.

Is your sample a liquid or a solid?

If your sample is a liquid (i.e. wastewater or potable water) then your best options would be CVAA or CVAF.

What are your detection limit requirements?

CVAA will provide a practical and secured detection limit of about 1 ppt and CVAF will provide a detection limit of about 0.1 ppt (or as low as 0.01 ppt with gold amalgamation).

Unless you have a preference for CVAA, I would recommend CVAF. Its superior detection limits will allow you to report to lower D.L. As well, a wide range of coverage will minimize your need for many sample dilutions.

To digest or not to digest?

Direct analysis by thermal decomposition, allows you to skip sample digestion needed for CVAA or CVAF. But, if your lab is already set up for sample digestion, the higher capital cost of direct analysis relative to CVAA (or CVAF), will be a factor.

Other factors, such as the sample homogeneity or volatility may be important considerations. Direct analysis is limited to a relatively small quantity of sample (about 1 g), nonhomogeneous samples may be best dealt with by digesting a larger quantity of sample followed by analysis using CVAA or CVAF.

I took over the management of an established analytical laboratories a while ago, and the detection level required was only in the ppb level. They dealt with soil samples that were very coarse and contained rock-type large segments. The sample taken was about 2 g. A larger sample would have meant high matrix with high levels of several parameters that would have interfered during the ICP or AAS role. This did not allow for having a homogeneous sample representing the mother-source. If you were to take another sample, with perhaps a large 5g of a different rock in the soil, you would get another result. The soil had different types of crystallized muddy pieces, different rock material, flakes and more. You would need a sample that reflects an average representation at large. At the same time, they took a small sample – as I said, to avoid heavy matrix interference.

I had to change the methodology to take a large 50g sample, dissolve to a 2L final volume, and then, you would take a sample from that prepared solution. In this case, obviously the direct solid sample analysis approach would not be an option.

Analytical Trap: This is the gold trap that captures the mercury from a sample trap and subsequently releases that mercury into the analyzer. The mercury analyzer is a cold vapor atomic fluorescence spectrophotometer (CVAFS).

In the CVAFS, light from a small mercury vapor lamp is shined through a quartz flow cell that contains mercury in a stream of argon carrier gas. This light excites the mercury atoms which subsequently emit more light.

The amount of light emitted by the mercury is proportional to the amount of mercury passing through the cell. The light emitted by the mercury atoms passes through a filter at the wavelength of 254 nm, and into a photomultiplier tube (PMT) which converts the light into an electrical signal. The signal is received and plotted by a strip chart recorder or integrator.

Now, let's establish some basic facts and terminology:

The blank: The mercury that is associated with equipment and reagents must be measured; this quantity of mercury must subsequently be subtracted from the mercury measured in samples.

The blank value will obviously be the measured quantity of mercury from background levels, and may refer to the association with specific objects, such as a bubbler blank or a BrCl blank, or association with specific processes, such the traveling blank or a digestion blank.

You may also need a blank value that is the sum of background contamination from all potential sources such as trace level in reagents.

The blank graph value involves the process of measuring the quantity of mercury that is associated with apparatus,

reagents, treatment and interference in the fashion that the sample and the standard and QC were exposed to.
The process of thermally desorbing mercury from a gold trap is commonly called blanking. This is sometimes also referred to as analyzing a trap.

Blank Matrix: An aqueous mixture which may contain any combination of deionized water, HCl, $SnCl_2$, BrCl, NH_2OH, HCl ... etc. This is the solution out of which elemental mercury is purged.

Coil: Coils are made from nichrome wire. These coils are used to desorb mercury from gold traps. Variable autotransformers are used to apply a potential of approximately 10 VAC to the coils; this heats the gold traps desorbing elemental mercury.

FEP: A type of Teflon (Fluorinated Ethylene-Propylene).
PTFE: A type of Teflon (Poly Tetra Fluoro Ethylene).

Glass Wool: Fibrous strands of silanized glass. Plugs of glass wool are used to hold soda-lime chips in soda-lime traps.

Gold Trap: Gold traps are made from a 9 cm length of 4 mm I.D. quartz tube. The tubes contain approximately 1 g of gold coated beads held in place with silanized glass wool plugs.

A constriction in the quartz tube holds all the packing materials in place. Since elemental mercury forms an amalgam with gold, the gold coated bead traps are used to pre-concentrate mercury purged from aqueous samples.

Pre-reduce: The addition of $NH_2OHCHCl$ to a brominated sample. Chemically reduces excess BrCl which could damage the gold traps.

Purge: To pass N_2 bubbles through an aqueous matrix to remove and trap the elemental mercury contained within that matrix.

Regulator Units: These consist of the single or two stage regulators on the N2 and Ar cylinders and any needle valves and other brass or Teflon fittings connected to these regulators.

Soda-lime Trap: Soda-lime traps are made from 10 cm lengths of 0.9 cm I.D. Teflon tubing.
They are packed with non-indicating, 8-14 mesh, reagent grade soda-lime; the ends are plugged with glass wool. These traps neutralize acid fumes and trap water vapor during the purging process.

Sample Traps: The gold traps that are first attached to the purging apparatus to capture elemental mercury and then connected to the analytical system for mercury measurement. These traps are connected to the soda-lime traps during bubbling and to the argon line and analytical trap during blanking.

Scrubbing Traps: The gold traps attached to the purging and analytical systems that are intended to remove contaminant mercury from the N2 and Ar gases.

Tenax TA: A porous polymer based on 2,6-diphenyl-p-phylene oxide used to trap matrixinterferents associated with some tributary samples. Placed between the gold trap and soda lime trap.

Timer Controller: A device used to switch coils and fans on and off at appropriate times during processing a gold trap blank.

Total Mercury: The sum of all the different chemical forms of mercury (includes inorganic and organic forms.)

A passing comment:

In the late 1990s I was the head analytical chemist of a major commercial analytical laboratory. Some of the samples were industrial sludge or waist solutions or solid samples of rock or metals, with a high level Hg, in the percentage range.

I want to include a note of experience here that might of benefit to lab technicians and junior chemists. Actually, to my surprise, experienced analytical chemists were just as guilty of getting into the routine, machine-like performance that often becomes a repeat procedure done without thinking. This is not to offend or insult anyone, but, if I've seen it done by chemists with a Masters degree and a decade experience, it might be done elsewhere by others.

It has to do with reporting a parameter in final percentage concentration of an element in a sample by weight. I noticed that what everyone was routinely obtaining a g/L result to convert it using specific gravity to a percent report as follows:

- They would record a sample weight.
- Digest it and prepare the sample for AAS to a certain volume, e.g. 500 ml or 1L flask.
- obtain the concentration of the parameter in solution, in ppm result, using AAS flame. From there, they have the result in g/L
- Take another sample, measure a specific volume, e.g. pipette 25 ml, then weigh that known volume sample, use the weight and the volume to get the specific gravity, and from there, convert the g/l to %.

I stood there unable to believe my eyes.
I called a junior chemist and I told him, you didn't need the spec. grv. in order to convert the ppm in solution to a final result in % in sample. He was confused, he repeated: I need to report it in % by weigh, not volumetrically g / L.
"Yes, I know." I said. This was the remainder of the conversation:

I : when you say ppm, what is that mean to you?
He: part per million
I: yes, what it that mean as far as quantitative reporting?
He: mg per liter
I: good. So, you have a known weight sample, let's say 0.5 g. You dissolve it in 1L and the AAS result is, say, 5 ppm in solution. So you have 5 mg per L. Well, if you have a liter, then you have 5 mg IN THE SAMPLE ITSELF.

5 divided by sample weight in mg X 100 is your percentage right there!

If your sample was made up to 0.5L then 5 ppm in solution = 2.5 mg   If you had 2L, you have 10 mg in the sample. If you had 150 ml, it's 5 ppm X 150 ÷ 1000 = mg in the sample. If you prepared "x" mg sample in 1L and AAS reading required 1:100 dilution and showed 5 ppm in final actual AAS run diluted sample, then it's 5 ppm X 100. Etc.   mg in sample ÷ sample weight in mg X 100 = final report of % by weight.

Clearly you don't need to calculate density or specific gravity. If you do, you'll introduce the additional margin of error caused by the volumetric pipette.

I found that every now and then I ran into a technician or a chemist who would do something like that, simply out of habit without stopping to think. The complexity of the possible formations of, for example, salts, during mercury lab work, and the need for a much higher level of mental concentration and "thinking" than that demonstrated in the preceding simple case, is imperative to achieve a successful and safe analysis.

Notes involving possible complications, side effects and compounds that may develop during the sample digestion or treatment:

- Controversy has existed, based on the vapour densities of mercurous compounds and the tendancy to disassociate, as to whether mercurous salts should be formulated as containing the univalent ion Hg or with the divalent ion Hg_2 containing two atoms of mercury. The equilibrium between mercurous and mercuric ions in contact with metallic mercury eliminates the univalent Hg ion – although this did not exclude complexes of the type $\{Hg_2CL\}$. Lab work on the conductivity of solutions of mercurous nitrate supports Hg_2 ion.

- mercury fluoride can be introduced from mercurous oxide or carbonate and hydrofluroric acid. high acidity will crystallize a colourless monoclinic acid salt $HgF,2HF,2H_2O$. A chemist approached me once asking about the appearance of a residue that was proven to be mercuric fluoride. This is because mercurous fluoride decomposes with heat.

- In one consultation case, a yellowish powder appeared, and it was proven to be mercurous chloride, which is produced for pharmaceutical use under the commercial name "calomel." This occurred during lab wet analysis without the chemist being aware of it. Evaporating sulphuric acid with excess mercury, creates mercurous sulphate, which, with heat, and in the presence of salt or by precipitating mercurous nitrate solution via nitric acid, is the explanation.
In fact, the preceding is a pharmaceutical procedure concluded with this step:

$$4Hg + 4HCL + O_2 \rightarrow 2Hg_2CL_2 + 2H_2O$$

- Organic compounds of mercury, are known in which mercury is bound directly to oxygen, nitrogen, sulphur. In oxygen-linked compounds, the mercury generally behaves as in inorganic salts. Compounds containing a N-Hg link will be formed if you allow for a reaction of mercuric salts or alkyl and arylmercuri-compounds with amines or amides.

- Here is a common lab occurrence:
  N-Mercurianiline $(C_6H_5NH)_2Hg$, an unstable white crystal formed as a reaction between mercuric chloride and sodium hydroxide on aniline.

- Mercury acetamide $(CH_3.CONH)_2Hg$ is produced if you heat acetamide with mercuric oxide. The N-Hg link is easily broken in acids, hydrogen sulphide or potassium iodide. Mercury succinimide has been used medicinally.

- Another note to remember is the organic compounds with a C-Hg link. They are common in samples of products used as antiseptic and fungicides products used in disinfection of seeds. I ran into several samples in labs that were aliphatic or heterocyclic. Mercury dialkyls R_2Hg is an example resulting from interaction of an alkyl halide with sodium amalgam in the presence of ethyl acetate. Also via a reaction of a mercuric salt and organo-zinc or organo-magnesium compounds.
  Here is an example of a possible complication:

  $R_2Hg + Br_2 = RBr + RHgBr$

  Also, the action of acids or mercuric chloride on mercury dialkyls, as in this case:

  $(CH_3)_2Hg + HCL \rightarrow CH_4 + CH_3HgCL$

$$(CH_3)_2Hg + CH_3.CO_2H \rightarrow$$
$$CH_4 + CH_3.CO_2HgCH_3$$

$$HgEt_2 + HgCL_2 \rightarrow 2EtHgCL$$

- Mercury diaryls are formed from a halogeno-compound such as bromobenzene, with sodium amalgam. It also formed by action of sodium stannite with mercuri-aryl acetate:

$$2PhHgX + Na_2SnO_2 + 2NaOH \rightarrow$$
$$HgPh_2 + Hg + 2NaX + Na_2SnO_3 + H_2O$$

- Hydrazine, which resembles ammonia in some aspects, forms with mercurous nitrate, a white crystalline mercurrous hydrazine nitrate $N_2H_4.2HgNO_3$ The bench chemist needs to pay attention to what I interpret or suspect it to be mercurous compounds breaking down into mercury and mercuric salts as soon as attempt is made to prepare ammonia derivatives. This would explain how mercurous chloride is slightly dissociated in this reaction:

$$2HgCL \leftrightarrow Hg+HgCL_2$$

and the equilibrium is slightly disturbed by the addition of ammonia, because it reacts the mercuric chloride forming an insoluble amide. This in turn reduces the conc. of the mercuric salt, and accordingly more is formed to restore equilibrium. This makes sence since that nearly all the mercurous ammonia compounds are black due to the presence of free mercury in a finely divided state.

This was observed when separating a mixture of mercurous and silver chlorides. Mercurrous chloride turned black with the addition of aq. ammonia. The black substance was long considered to be dimercuroammonium chloride, then, it was considered dimercuric amidochloride formed this way:

$$\begin{array}{c}\text{Hg} \text{—} \text{CL} \\ \text{Hg} \text{—} \text{CL}\end{array} + 2NH_3 \rightarrow \begin{array}{c}\text{Hg} \text{—} \text{CL} \\ \text{Hg} \text{—} \text{NH}_2\end{array} + NH_4CL$$

Since then chemists considered it a mixture of finely divided mercury and mercuric chloride. If you think about it, if a thin layer of the black precipitate is spread on glass and dried, it will lose weight in a few days and turn grayish-white colour. The loss in weight is due to the black substance, a mixture of free mercury and mercuric chloroamide CL-Hg-NH$_2$ and mercury volatilizes leaving behind CL – Hg – NH$_2$. This makes sense within this reaction:

2HgCL + 2NH$_3$ → Hg + CL.Hg.NH$_2$

The proof that the black precipitate contains metallic mercury is that if mercurous and silver chlorides are jointly precipitated and allowed to stand for some time in contact with ammonia, some metallic silver is formed due to the presence of free mercury as follows:

AgCL + Hg = Ag + HgCL

- Mercury slowly reduces aq. soln. of ferric chloride, and rapidly if the soln is acidified with hydrochloric acid:

FeCL$_3$ + Hg = HgCL + FeCL$_2$

- Sulphur dioxide oxidizes mercurous to mercuric chloride:

SO$_2$ + 4HCL + 4HgCL = 4HgCL$_2$ + 2H$_2$O + S

- It was reported that when calomel is heated with dry

alkai hydroxides, or the oxides or hydroxides of the alkaline earths, mercury, oxgen and the alkali chloride are formed. This is based on:

$$4HgCL + 4KOH = 4KCL + 2H_2O + O_2 + 4Hg$$

and with aq. soln. of the alkali hydroxides black mercurous oxide if formed:

$$2HgCL + 2KOH = Hg_2O + H_2O + 2KCL$$

- Hypophosphorous acid precipitates metallic mercury from soln of mercuric chloride, but if the mercuric salt is in excess, mercurous chloride is formed.

- Phosphorous acid precipitates mercurous chloride completely from soln of mercuric chloride. When the mercuric chloride is boiled with excess of phosphorous acid, the mercurous chloride is grey, and mercury is obtained only in the presence of free acid. the reaction occurs in two steps:

$$HgCL_2 + H_3PO_3 = HCL + HgCL + H_2PO_3$$

$$2H_2PO_3 + H_2O = H_3PO_3 + H_3PO_4$$

- When dry mercuric chloride is heated with the alkali hydroxides or carbonates, a sublimate of mercury and a residue of alkali chloride is formed. When hydroxide of the alkalies or alkaline earths is added to a soln of mercuric chloride, a reddish – brown precipitate of a basic chloride $Hg_3O_2CL_2$ or Hg_2OCL_2 is formed provided the mercuric salt be in excess:

$$2HgCL + 2KOH = 2KCL + H_2O + HgCL\_O\_HgCL$$

or:

3HgCL$_2$ + 4KOH = 4KCL + 2H$_2$O + Hg$_3$O$_2$CL$_2$

whereas if the alkali be in excess, yellow mercuric oxide is precipitated:

HgCL$_2$ + 2KOH = 2KCL + H$_2$O + HgO

The chemist has to consider the system acquiring a state of equilibrium as follows:

HgCL$_2$ + 2KOH ↔ 2KCL + HgO + H$_2$O

Understanding the possible complications in your test tube or beaker, based on the possible presence and interaction of many compounds in your sample, is essential for the bench chemist's success.

Total Mercury Analysis in Aqueous Samples

Working Reference Standard:

A solution with a known concentration of Hg2+.
This solution, mixed form more concentrated primary and secondary standards, is used to calibrate the instrument. This is similar to the application of Standard Addition approach applied to QC-MM "Matrix Match" and to duplicate samples, repeatedly inserted for QA purpose.
For example adding 2 ml with a value of additional 10 ug of mercury to the test tube of the duplicate "replicate" unknown sample, as well as to a 2nd QC.

Cautions

BrC is extremely corrosive, and its solution releases Cl2, Br, and BrCl fumes which, being toxic and extremely caustic, will cause severe damage to the respiratory system if inhaled. Flush affected areas with water and mild soap. A strong suction fume hood is required.

Soda-Lime: Can cause burns. Avoid contact with skin and eyes. Rinse affected areas with large amounts of water.

Total Mercury Standard: Mercury in the standard can damage the nervous system. Avoid contact with skin. The working standard contains up to 5% BrCl.

Preparing Water Samples for Analysis

- Digest the water samples
- Add excess BrCl to every water sample. This will be indicated by a persistent yellow colour. For tributary samples generally 5 to 8 ml per 500 ml flask are required. Record

the amount of BrCl added to the sample because mercury in any of its chemical forms is oxidized to Hg^{2+} by BrCl in the sample so the appropriate reagent blank can be subtracted.

Preparing the reagents

- Stannous Chloride

200 g of $SnCl_2$ is poured into 1 L Teflon bottle and then 100 mL of concentrated Tracepure HCl is added. Add analytical grade water to bring the volume to 1 L. The solution is purged with N_2 overnight

- Hydroxylamine Hydrochloride

300 g of NH_2OH-HCl crystals are poured into a clean Teflon 1 L bottle and then filled to 1L with analytical grade water. Cap the bottle and shake until all crystals are dissolved. For each 1 L of solution, add 1 mL of $SnCl_2$ and purge the solution overnight with N_2.

- Bromine Monochloride

Caution: all work must be done in the fume hood and wearing safety goggles.

8.6 g of KBr is poured in the 1L Teflon bottle (BrCl stock solution bottle) and the bottle is then filled with 800 ml of concentrated Tracepure HCl and a clean magnetic stir bar added. For approximately an hour the solution is stirred in the fume hood with a stir plate. After the hour is up, *very slowly* add the $KBrO_4$ crystals while stirring slowly. Add small amounts of crystals and only add more when fizzing has stopped. Then allow the solution to stir with the cap on loosely for another hour.

- Mercury Standards

Normally a stock mercury standard of 1000 mg/L is one of the SRF used. Pipette 100.0 FL of the stock solution and 5 mL of BrCl solution into the volumetric flask and dilute to 100.0 ml with analytical grade-Q water. After mixing, the

solution is poured into a clean 125 mL Teflon bottle and labeled with Hg₂, ID of stock solution, the date, and your initials. This can be stored in a refrigerator for up to one year.
Standard Operating Procedure for the Lake sample, using Cold Vapor Atomic Fluorescence Spectrophotometer (CVAFS)

- Mercury is measured using HgNO₃ standard from two sources such as Fisher Scientific and Varian.
- Calibrate the CVAFS with a four point curve spanning the anticipated range of sample signals.

- This calibration is achieved using a vapor mercury standard. As mercury is a liquid at room temperature it has a well-defined vapor pressure, which is temperature dependent. A temperature-controlled sealed container containing liquid mercury will therefore have a defined air concentration of mercury. This mercury-saturated air can be sub-sampled via a septum port using a gas tight chromatography syringe. Known aliquots of the air are injected into the gas stream of the dual gold trap-CVAFS system, and trapped on the gold column just prior to the instrumentation part. Heating the column releases the Hg into the instrument allowing quantification of the injected concentration.

Interferences

Interferences have been reported for waters containing sulfide, chloride, copper and tellurium.
Organic compounds which have broad band UV absorbance (around 253.7 nm) will show conflict.
Volatile materials (e.g. chlorine) which absorb at 253.7 nm will cause a positive interference.
In order to remove any interfering volatile materials, addition of stannous chloride solution should be followed by swirling of the digestion vessel before pouring the contents into autosampler vials.

Low level mercury sample preparation, digestion, and analysis may be subject to environmental contamination if performed in areas with ambient backgrounds where mercury was previously employed as an analytical reagent in analyses such as total Kjeldahl nitrogen (TKN) or chemical oxygen demand (COD).

APPARATUS AND MATERIALS

The following case documentation is an illustration performed by a fellow chemist. I had produced a very similar paper in 1995, but I used a Vaian instrument. My paper was published online early 1995. In this case, Leeman PS200 Mercury Analyzer is utilized – which includes the optical cell, mercury lamp, peristaltic pump, drying tube, sample and reagent tubing and Leeman PS200 software.
Reagent water – which here refers to ASTM Type II grade water.
Hydrochloric Acid (conc. HCl), trace metals grade or better.
Stannous Chloride," $SnCl_2$, ACS Reagent Grade or better," suitable for mercury determination.

10% $SnCl_2$ in 10% HCl. Dissolve 100 g of stannous chloride (7.3.3) in 100 ml HCl and dilute to 1L with reagent water. Place a stirring bar into the solution.
Record the preparation of this reagent in the Inorganic Standards and Reagents Preparation Log Book.
Argon gas, ICP grade.

Magnesium Perchlorate, powder or granular.
10% HCl. Add 200 mL HCl (7.3.2) to approximately 1.5 liters reagent water and dilute to 2L with reagent water. Record the preparation of this reagent in the Inorganic Standards and Reagents Preparation Log Book.

QUALITY CONTROL PROCEDURES

The QC programme consists of an initial demonstration of laboratory capability, and the periodic analysis of laboratory reagent blanks, fortified blanks, QCS samples and other laboratory solutions as a QC – QA performance monitoring. The laboratory must maintain performance records that define the quality of the data that are generated.

Initial Demonstration of Performance - Prior to use in routine analysis, the method must be characterized in terms of calibration linearity and MDL.
LDR - Absorbance vs. mercury concentration should be a linear function over a limited concentration range. Prior to using this method for routine analysis, the calibration range must be characterized and linearity demonstrated.

The characterization must be verified routinely or whenever a significant change in instrument response is observed or expected (e.g, a new optical cell is installed). The initial characterization is performed by analyzing a calibration blank and a series of calibration standards with a concentration range spanning the expected sample concentration range.

The resultant absorbance vs. concentration data are plotted and examined for linearity using professional judgment.
A calibration line is calculated by linear regression, using at least 4 data points. To be considered linear, the correlation coefficient must exceed 0.995.

Note:
When evaluating the adverse health effects of mercury and clarifying its dynamics and pathways in humans and in the environment, in addition to performing a quantitative analysis for total mercury, one should also perform separate quantitative analyses for methylmercury and inorganic mercury. This paper provides you with a description of sampling followed by descriptions of the analytical methods for total mercury and methylmercury for each target sample.

Sampling:
> Environmental samples
> Biological samples (e.g. fish and shellfish)

Water pollution caused by methylmercury can be monitored conveniently by measuring the bioaccumulation of mercury in fish. Further, monitoring the mercury in fish and shellfish eaten most often by people in a particular region is a suitable means of evaluating human exposure levels, because human exposure to methylmercury occurs mostly through the consumption of fish and shellfish.

According to Japan's Food Hygiene Act, the provisional regulatory standard for mercury in fish and shellfish is 0.4 mg/kg (wet weight) as total mercury. The background level of total mercury in fish and shellfish is considered to be 0.01-0.1 mg/kg (wet weight).

Water:
When the source of contamination is directly connected to a river, lake, marsh, or ocean, or when contamination is expected to have spread from a river to a lake, marsh, or ocean, take water samples from the various areas. Use a Bandon water sampler or a similar approach to collect the water samples, preferably at 20-30 cm below the surface. Take great care to prevent bottom sediment from entering water samples collected near the bottom. In principle, collect seawater samples at high tide and avoid windy or rainy days. For total mercury, Japan's Effluent Standard and Environmental Quality Standard are 0.005 mg/l (ppm) and 0.0005 mg/l, respectively, according to Environment Agency Notice No. 64, based on the Water Pollution Control Law in Japan. For alkyl mercury, both the Effluent Standard and Environmental Quality Standard state that it must not be detectable at the detection limit of 0.0005 mg/L when analyzed with the official methods provided by the Water Pollution Control Law.

When a contamination accident causes a sudden release of wastewater containing high levels of mercury, the levels of total mercury and alkyl mercury must be evaluated according to these official standards. The background levels of total mercury are generally 0.5-3 ng/L (ppt) for ocean water, 2-15 ng/L for shore water, and 1-3 ng/L for rivers and lakes.

Sediment/soil:

When collecting soil samples, vary the frequency of sample collection depending on the plane position of the mercury pollution source and the extent of the suspected contamination. While various methods for collecting soil samples from the site have been proposed, Japan's Soil Pollution Countermeasures Law provides detailed descriptions of the methods for collecting soil samples. These methods served as the basis for the sampling methods provided in this paper.

Plants:

Plants normally exhibit little biological magnification of heavy metals and therefore are not suitable for evaluating contamination. However, lichens have various properties that make them suitable as a biological indicator of air pollutants. As with other rootless air plants, they absorb nutrients directly from the air, accumulate metals efficiently, and exhibit resistance to high concentrations of metals in their tissues.

Lichens are widely distributed geographically, making them suitable for not only domestic but also international evaluations of air pollution. In fact, lichens (including the Parmelia and Usunera species) have often been used in research to evaluate air pollution caused by mercury and various other heavy metal pollutants, and Garty (2001) has reported a related review article. Lichens, which usually grow on trees or branches, are collected, washed well with water, cleaned of wood pieces and dust, and air-dried to make a sample. For mercury analysis, place a few grams of the sample in a vial and cut it into pieces with dissection scissors.

Atmosphere/air:
Air samples are collected when mercury pollution is believed to be present in the atmosphere or indoor environment. Since mercury concentrations in the atmosphere vary greatly, sampling points must be selected in order to clarify the mercury distribution with consideration given to prevailing winds and the distance from the contamination source. To collect mercury in the atmosphere or in the indoor air, place an absorbing solution comprising 20 ml of 0.1% potassium permanganate and 1N sulfuric acid in a bubbler. To sample the air, use a suction pump to draw the air into the absorbing solution from the sampling point at a flow rate of 1 L/min. Because commercially available potassium permanganate often contains mercury:

1- use the same solution to prepare a blank QC to calculate the amount of Hg inserted.
2- dissolve it in 1N sulfuric acid and boil it to generate a precipitate of MnO_2. Cool and filter it for use as an absorbing solution. This procedure can remove all mercury content and render the absorbance of the blank test solution almost zero. The absorbing solution obtained in this manner can collect mercury vapor efficiently, and one bubbler containing absorbing solution is usually sufficient. Personally, I would still use that solution for a comparative blank QC, just to be certain.

Note: If the absorbing solution has evaporated and decreased in volume after the air samples have been drawn, top up the absorbing solution to a fixed volume to make a test sample. Separately from this sample, take two identical volumes of absorbing solution that have not been aerated. Set aside one volume as a blank test solution; to the other, add a fixed volume of inorganic mercury (II) standard solution to create a standard test solution.

At measurement, add 10% hydroxylamine hydrochloride dropwise to decolorize the potassium permanganate;

determine the mercury concentration in the sample test solution by cold vapor atomic absorption spectrometry, as with the other samples. Using the volume of air collected, calculate the mercury concentration in the air sample. This method can be widely applied to tests of the environmental atmosphere, the air of a work environment, and the gas discharged from an emission source or the like.

For mercury in the atmosphere, the Japan Society of Industrial Health recommends a concentration of 0.025 mg/m3 as the permissible concentration of mercury vapor in the work environment.

Human samples:

Hair:
The mercury concentration in hair is often used as a biomarker for methylmercury exposure because it reflects the concentration in the blood at the time the hair was formed. At the same time, a hair sample provides a simple and noninvasive sampling method as well as a storage method offering good sample preservation. The mercury concentration in hair is generally detected at 250-300 times the blood concentration. Since the hair grows at a rate of roughly 1 cm per month, evaluation of past exposure is possible. However, the mercury concentration in hair can increase as a result of adhesion of external mercury vapor and inorganic mercury, decrease as a result of hair treatments such as permanents, and be influenced by the sample collection site.

In cases of no exposure to external inorganic mercury or mercury vapor, almost all mercury in hair is in the form of methylmercury; therefore, the level of methylmercury exposure from diet can be evaluated by measuring total mercury. Among the general population in Japan, the mercury concentration in hair is in the range of 1-5 ppm and seldom exceeds 10 ppm.

Blood:
For people who eat large quantities of fish and shellfish, the mercury concentration ratio of red blood cells to plasma (serum) is approximately 10:1, and most mercury contained in the red blood cells is in the form of methylmercury; therefore, the methylmercury exposure can be evaluated by measuring total mercury in blood. It is believed that 50% of inorganic mercury is present in the plasma and the mercury concentration in the plasma increases in relation to the amount of inorganic mercury accumulated in the kidneys. Thus, the exposure to inorganic mercury/mercury vapor can be evaluated by measuring the total mercury in plasma.

A blood sample in the range of several milliliters is collected into an injection tube already containing an anticoagulant (heparin) and transferred into a sealed container. The sample is then centrifuged at 3,000 rpm for 10 minutes to separate the red blood cells from plasma. Samples to be stored for a long period of time should be frozen. While the mercury concentration in the blood of the general population in Japan generally does not exceed 40 ng/g, people whose diet is rich in fish sometimes have higher values.

Urine:
Most mercury present in the urine is in the form of inorganic mercury. The mercury concentration in the urine increases in relation to the level of inorganic mercury accumulated in the kidneys. Accordingly, the total mercury value in the urine is an important biomarker for evaluating inorganic mercury/mercury vapor exposure. On the other hand, leaking of methylmercury into urine may occur in those with renal disease.
Since the mercury concentration in the urine also varies with the excretion rate, it is necessary to correct it with the creatinine concentration in the urine or to collect the urine sample at a designated time.
Generally, as in usual urinalysis, 50-100 ml of urine is collected as a sample in a paper cup in the early morning. The sample is then stored under refrigeration in a polyethylene

container. Samples to be stored for more than one month should be frozen. Since urine contains many inorganic salts, even fresh urine may generate precipitate. Thus, the sample must be homogenized by shaking before analysis. A method also exists where the solubility of the salts is increased by lowering the pH of the urine sample by adding a small amount of hydrochloric acid.

Take steps to ensure that microorganisms do not proliferate, as they may cause inorganic mercury to reduce to mercury vapor, which will escape and be lost.

It is estimated that the average mercury level in the urine of the general population in a region without any particular mercury exposure is less than 10 ng/ml.

Of course any additives introduced to the urine sample, even analytical grade HCL, would be added to a blank QC prepared with the same quantities, and would undergo the same treatment steps.

Umbilical cord:

When food containing methylmercury is ingested by the mother during pregnancy, the methylmercury easily passes through the placenta, transferring from the mother's body to the fetus and exposing the fetus to the methylmercury. One prime example of this was the occurrence of many fetal cases of Minamata disease accompanied by severe cerebral palsy-like symptoms.

This was caused by pregnant mothers' ingestion of fish and shellfish highly contaminated with methylmercury when Minamata disease was prevalent in Japan. Mercury in the umbilical cord is used as a suitable biomarker of methylmercury exposure during the fetal period because the mercury concentration in the umbilical cord is highly correlated with that in umbilical cord blood, and most mercury occurring in the umbilical cord is in the form of methylmercury.

Several centimeters of the fetus side of the umbilical cord are collected at delivery and washed with physiological saline to remove the blood. The sample is stored frozen until the time of analysis. The stored umbilical cord is soaked in water to moisten it, washed with water to remove the blood and other adhering substances, and air-dried to prepare the sample for analysis.

The concentration of methylmercury in the umbilical cord in the general population in Japan is considered to be around 0.1 µg/g (dry weight). It has been reported that the methylmercury concentration in the umbilical cord can be as high as several µg/g (dry weight) in children born during the Minamata disease outbreak.

Lab notes:

Analytical Method for Total Mercury:

The conventional methods for measuring total mercury include absorption spectrometry (dithizone colorimetry), neutron activation analysis, and cold vapor atomic absorption spectrometry.

In absorption spectrometry, dithizone forms a complex with the metal ions and produces a colored organic solution. The color intensity varies with the mercury concentration. Although this method has been used historically because of the simplicity of the procedures, its use declined greatly with the introduction of highly sensitive atomic absorption spectrometry.

In **neutron activation analysis**, thermal neutrons in a nuclear reactor are irradiated and gamma radiation from generated 197Hg is measured for comparative quantification with the standard sample. This enables nondestructive analysis in which the sample is analyzed directly without any pre-treatment such as concentration, and it is highly precise and sensitive. However, it is not used frequently due to its high cost, the need for a nuclear reactor, and the need for an expensive counting apparatus, not to mention the safety requirements for handling radioactive materials.

In **cold vapor atomic absorption spectrometry**, mercury is converted into elemental mercury vapor, which is introduced into an absorption cell and the absorption measured at 253.7 nm for determination of the quantity. It is a much more sensitive method as compared with conventional flame atomic absorption spectrometry.

Other advantages include its ability to measure mercury in the samples with a UV spectrophotometer or a simple mercury lamp. It is roughly classified into the reduction/aeration procedure and the sample combustion procedure according to the generation mode for mercury in the elemental form.

The former involves wet digestion with a mixture of strong acids followed by the addition of a reducing agent to generate elemental mercury vapor (Hg0). In the latter, elemental mercury vapor (Hg0) is generated through direct combustion of the sample to be analyzed.

Herein we describe -among these highly sensitive analytical methods- a method involving wet digestion, reduction and cold vapor atomic absorption spectrometry (CVAAS) (the circulation-open air flow system), which offers substantial improvements over the conventional method.

Determination by the wet digestion/reduction/cold vapor atomic absorption spectrometry (CVAAS) (circulation-open air flow system).

Principle:

The present method involving reduction and cold vapor atomic absorption spectrometry (CVAAS) (circulation-open air flow system) is, in principle, similar to the conventional circulation system in that the method includes the following:

Reduction of Hg_2+ ions in the sample test solution with stannous chloride to generate elemental mercury vapor ($Hg0$); and the introduction of mercury vapor into the photo-absorption cell for the measurement of absorbance at 253.7 nm.

However, unlike the conventional closed system in which the elemental mercury vapor generated is continuously circulated with a diaphragm pump through a reaction vessel, a U-shaped tube packed with a drying agent, and the photo-absorption cell, the present method uses a circulation-open air flow system.

The apparatus constitutes a closed system and comprises a diaphragm pump, reaction vessel, acid gas trap, moisture trap (ice bath), and a 4-way cock. During its operation, the elemental vapor generated by the addition of stannous chloride is circulated via the 4-way cock at a flow rate of 1-1.5 L/min. for 30 seconds to homogenize the concentration in the gas phase. The 4-way cock is then rotated by 90° to introduce the gas phase into the photo-absorption cell all at once.
The measurement is completed within one minute per sample with this apparatus, which can measure even 0.1 ng of mercury with high accuracy.

Additionally, in the method for preparing the sample test solution for the present method, the conventional wet digestion method is improved by the use of a 50-ml flask with a long neck (at least 10 cm), such as a thick-walled volumetric flask with a ground glass stopper, as well as a mixed acid system with an increased rate of sulfuric acid, HNO_3 - $HClO_4$ - H_2SO_4 (1+1+5), that already contains perchloric acid, for the sample digestion.

It is a simple method where the sample is subjected to wet digestion on a hot plate at 200-230°C for 30 minutes and cooled followed by topping up to a fixed volume with water. This method can be applied directly to the digestion of biological samples including hair, blood, and fish as well as various solid samples such as sediment and soil.

Reagents:

- HNO_3 - $HClO_4$ (1+1): Mix 100 ml of perchloric acid (for measurement of toxic metals) into 100 ml of nitric acid (for measurement of toxic metals). (Store in a cool dark place.)

- H_2SO_4: Sulfuric acid (for measurement of toxic metals)
- Distilled deionized water, in a clean glass container.

- HCl: Hydrochloric acid (analytical grade)

- 10% $SnCl_2$ solution: Dissolve 10 g of tin (II) chloride dihydrate (analytical grade), $SnCl_2 \cdot 2H_2O$, in 9 ml of HCl and dilute to 100 ml with distilled water. Aerate with N_2 gas (100 ml/min., 20-30 minutes) to expel any mercury from the solution.

- 5N NaOH: Dissolve 20 g of sodium hydroxide (analytical grade) in distilled water to make a final volume of 100 ml.

- 0.1N NaOH: Dilute 5 N NaOH 50-fold with distilled water.

- 0.1% L-cysteine solution: Dissolve 10 mg of L-cysteine hydrochloride, $HSCH_2CH(NH_2)COOH \cdot HCl \cdot H_2O$, in 10 ml of 0.1N NaOH. (Prepare a fresh solution for each analysis.)

- Methylmercury standard solution1: Weigh out 12.5 mg of CH_3HgCl (authentic standard) in a 100-ml volumetric flask, dissolve in toluene to make a final volume of 100 ml, and store as stock solution. Dilute the stock solution 100-fold with toluene to obtain a methylmercury standard solution. One ml of this solution contains 1.0 µg of Hg.

- Methylmercury-cysteine solution: Transfer 0.5 ml of the methylmercury standard solution and 5 ml of the 0.1% L-cysteine solution into a 10-ml conical centrifuge tube with a glass stopper. Shake for 3 minutes with a shaker to extract methylmercury into the aqueous phase. Centrifuge at 1,200 rpm for 3 minutes and draw off and discard the organic phase (upper phase). Seal the tube and store in a cool dark place. (Prepare a fresh solution monthly). One ml of this solution contains 0.1 µg of Hg.

- 1N H_2SO_4: Gradually add 30 ml of sulfuric acid (for measurement of toxic metals) to distilled water to make a final volume of 1,000 ml.
- 1% acidic $KMnO_4$ solution for collecting mercury: Dissolve 1 g of potassium permanganate (analytical grade) in 100 ml of 1N H_2SO_4.

- 0.5% $KMnO_4$ solution: Dissolve 0.5 g of potassium permanganate (analytical grade) in distilled water to make a final volume of 100 ml.

- Toluene: $C_6H_5CH_3$ (reagent grade for residual pesticide analyses)

Lab notes

Instruments and equipment:

Mercury analyzer Model Hg-201 Semi-Automated Mercury Analyzer (Sanso Seisakusho Co.)
Hot plate: Capable of attaining a surface temperature of 250°C
Sample digestion flask3: 50-ml thick-walled volumetric flask made of Pyrex (150 mm total height, 13 mm inlet diameter)
Volumetric flasks: 10, 100, and 1,000 ml
Measuring pipettes: 0.2, 0.5, 1.5, and 10 ml
Vial: 20-ml scintillation vial centrifuge.

10-ml conical centrifuge tube with glass stopper: 16.5 mm in diameter × 100 mm in length
Dissection scissors
Multi-flow meter: V4-type flow meter multi-kit (Kojima Instruments Inc.)
Reciprocal shaker

Note: Prior to use, thoroughly wash all laboratory glassware and sample containers to be used in the analysis with a 0.5% $KMnO_4$ solution. Rinse with water until the color of the $KMnO_4$ solution is no longer visible.

Preparation of sample test solution:

Precisely weigh out a homogenized sample (0.5 g maximum of wet weight) and place at the bottom of a sample digestion flask. (For dry samples such as umbilical cord, weigh out precisely 0.1 g and add 0.5 ml of distilled water to moisten beforehand.) Add 1 ml of distilled water, 2 ml of HNO_3-$HClO_4$ (1+1), and 5 ml of H_2SO_4 in turn and heat on a hot plate at 200-230°C for 30 minutes Allow to cool, add distilled water to make a fixed volume of 50 ml, mix well, and use the resulting solution as the sample test solution.

For urine samples, add 2 ml of HNO_3-$HClO_4$ (1+1) and 5 ml of H_2SO_4 to a sample digestion flask beforehand. Gradually add a fixed volume (usually 2 ml) of the urine sample while stirring slowly. Heat and treat it in a manner similar to the above procedure to prepare the sample test solution4.

Separately, transfer 0 and 1.0 ml of methylmercury-cysteine solution (0.10 µg Hg/ml) into two sample digestion flasks (corresponding to 0 and 0.10 µg Hg). Add 1 ml of distilled water only to the former (the blank) followed by 2 ml of HNO_3-$HClO_4$ (1+1) and 5 ml of H_2SO_4 in turn. Follow the same steps as indicated above for preparation of the sample test solutions to make a blank test solution and a standard test solution for the measurement of total mercury.

Test procedures and calculations

Gently transfer fixed volumes V ml (usually 5 ml, to a maximum of 10 ml) of each of the blank test solution, the standard test solution, and the sample test solution into the reaction vessel of the mercury analyzer and apply the stopper. Add 1 ml of 10% $SnCl_2$ solution with the accessory dispenser and push the start button.

The diaphragm pump will run and the generated elemental mercury vapor will be circulated through the 4-way cock between the reaction vessel and the acidic gas trap for 30 seconds to homogenize the concentration in the gas phase, while the acidic gas generated fro

After 30 seconds, the 4-way cock will turn automatically by 90°, allowing the introduction of mercury vapor into the photo-absorption cell through an ice bath for measurement of the absorbance. The readings of the recorder will increase sharply and decrease with a sharp peak. When the recorder reading begins to decrease, open the cock on the lower part of the reaction vessel to discard the solution inside, close it again, and allow it to aerate until it returns to the baseline.

The peak heights (mm) obtained after measurement of fixed volumes V ml of each of the blank, the standard, and the sample test solutions (or their diluted solutions) for the total mercury analysis are labeled Pbl, Pstd, and Ps, respectively. The total mercury concentration in the sample is calculated.

Total mercury concentration in the sample (µg/g) = 0.10 µg × (Ps–Pbl)/(Pstd–Pbl) × dilution factor × 1/sample weight (g)

For blood and urine samples, the mercury concentrations are usually expressed in ng/g and ng/ml, respectively, and thus calculated with the following formula:

Total mercury concentration in blood or urine (ng/g or ml) = 100 ng × (Ps–Pbl)/(Pstd–Pbl) × dilution factor × 1/sample amount (g or ml)

Procedural Notes

Although a standard solution of inorganic mercury (II) is commonly used as a standard solution for the analysis of total mercury in the sample, the present method uses a methylmercury-cysteine solution as the standard solution, the same used for methylmercury analysis.

Similarly to the samples, it is subjected to wet digestion in order to make a standard test solution for total mercury measurement. This is an effort to avoid measurement errors caused by the use of a different standard solution, because most mercury contained in fish and shellfish is in the form of methylmercury, and the total mercury as well as methylmercury in the sample are commonly measured at the same time. Methylmercury in the organic solvent is extremely stable.

Even 1 ppm of methylmercury in a toluene solution can be used for several years if stored frozen to prevent volatilization of the solvent. When the preparation of a standard solution for total mercury measurement using the present method unavoidably requires the use of an inorganic mercury (II) standard, the following method is recommended for its stability, good storage characteristics, and other advantages. Inorganic mercury standard solution:

Weigh out 13.5 mg of mercury (II) chloride (standard) in a 100-ml volumetric flask, dissolve in 4 ml of HNO_3-$HClO_4$ (1+1) and 10 ml of H_2SO_4 added in turn, and top up to the mark with distilled water to make a stock mercury solution (1 ml of the stock mercury solution = 100 μg Hg). The stock mercury solution obtained in such way will be stable for several years if sealed and stored in a cool dark place.

At every use, the stock solution is diluted 1,000 times with the above blank test solution to make a mercury standard solution (1 ml of this solution = 0.10 μg Hg). In addition, when a commercially available standard solution is used, the blank test solution is similarly used to dilute it appropriately.

The automated apparatus based on this principle is commercially available as a Model Hg-201 Semi-automated Mercury Analyzer (Sanso Seisakusho Co. Ltd., Tokyo, Japan).

The equilibrium concentration between the aqueous phase and the gas phase of reduced and vaporized mercury vapor may differ depending on the acid concentration and volume of the sample test solution at measurement. Therefore, the blank test solution is used for dilution of the sample test solution and both the sample test solution and the standard test solution are measured under the same conditions in every respect (acid concentration and volume).

In atomic absorption spectrometry, the multi-point calibration curve method is not always required because the linear range of the calibration curve is wide. Therefore, a one-point calibration curve method is often used as well. Moreover, in addition to the blank solution, choose the most suitable concentration of the standard test solution from, for example, 0.02, 0.05, or 0.10 µg Hg/50 ml for a total mercury measurement with a peak height close to that of the sample test solution. In this case, use the same volume of both the standard test solution and sample test solution during the measurements. This will facilitate quantification.

A passing comment:

On the subject of calculating a concentration value on a "calibration" & a "run" graph:

Sometimes you might need to manually calculate a value of an unknown sample, correctly located on the graph somewhere between two QCs before it and after it.

The instrument will do that computing for you, but, there are cases when you may want to zoom in mathematically on that critical sample to confirm its value. There are cases when there is no other way of obtaining the value of the sample other than by an old fashion manually graph paper.

This is specially true with volumetric bench analysis where a 4 or 5 points graph is drawn based on known value responses to standard QCs. The unknown sample value is then concluded based on it placement on the X-Y graph, or its location on the curve or straight line between two QC points.

Let's look at the following two graphs.

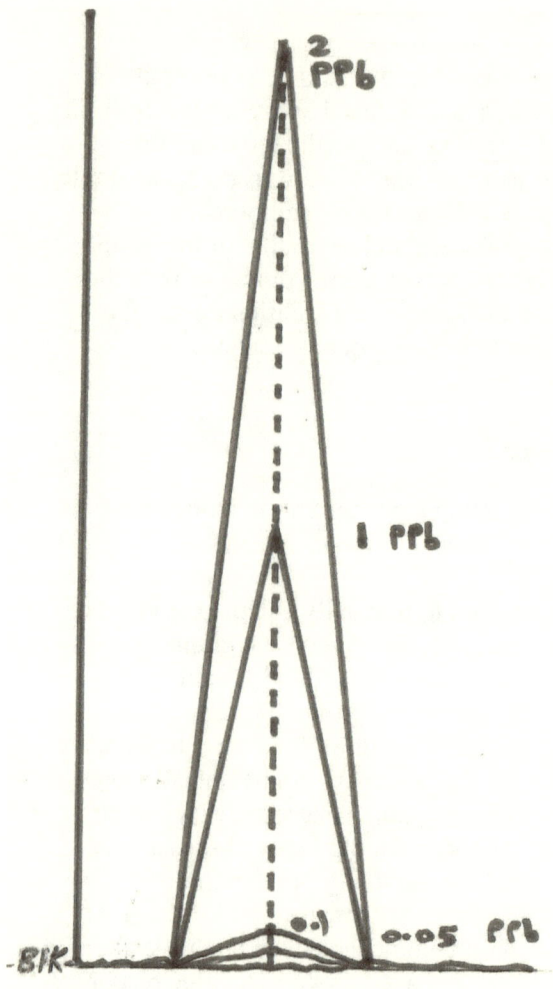

In this case, the calibration graph is "0" blank, 1 ppb and 2 ppb. As well, 0.1 ppb and 0.05 ppb were entered to monitor performance at low level near the BDL area.

It is assumed of course that other QCs including matrix match known samples (standards) have been run and are located on the dotted line graph above between 0 to 2 ppb.

Hydrargyrum analysis – Dr. Paul Gouda

Lets us look at the same calibration curve from a different angle.

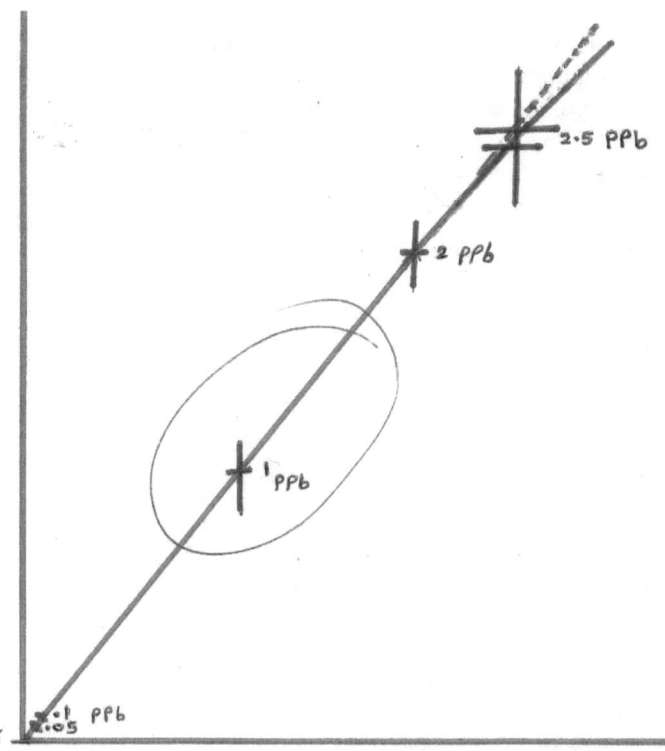

As shown, the target area, which is circled, is the area within which you would plan to have your samples, QCs, & Calib. Std reruns.   MM = matrix match.

This kind of accuracy and instrumental sensitivity demonstrated in a perfect graph (a straight line), means that between 0.5 and 2 ppb in this case you basically don't need a chart or a graph paper, you don't need a graph even if the instrument provided readings only (e.g. comparative abs. values of calib. STDs, QCs & the sample located between higher and lower QCs.)

In other words, you can calculate the value of a graph point mathematically without the need for a graph paper.

In fact, if for example, you are working within 5 to 20 ppb concentration area, with consistent instrument readings (MM blank, Std., QC & sample) – and you have a blank within a low noise line, basically "0" value. You also have a low, diluted QC "MM":

Say 0.1 ppb with obvious response beyond the blank value or noise line, and you have for example (mere simplification):

50 = 0.1 ppb
1000 = 2 ppb
2500 = 5 ppb
5000 = 10 ppb
10,000 = 20 ppb

Between 2 to 20 ppb you have a secured and safe straight line. Any unknown sample falling within this area can be calculated without a graph.

You clearly have a perfect correlation cf. greater than 0.999 which is in reality the ideal analytical laboratory target.
In other words if 20 ppb read 9,993 or 5 ppb read 2,506, considering the fractional percentage of 1 ppb and your report expectancy and format, you still have a straight line.

This is how you would calculate your sample value:

Take two nearest QC readings, one before and one after the sample reading. It is assumed that these readings are corrected based on MM & baseline noise.

Meaning, if your sample read 3869:
You would eliminate all other values but 2500 & 5000 points. The nearest two QC values before and after the sample.

Hydrargyrum analysis – Dr. Paul Gouda

Between these two QC points you have 2500 graph points, which is 5000 – 2500 (the distance between 5 ppb & 10 ppb.)

2500 graph area points = 5 ppb area (the difference between two QCs 10 – 5 ppb)
2500 ÷ 5 = 500 graph area points per each 1 ppb
which is 50 graph points per each 0.1 ppb "which is consistent with low QC detection."

Now, mathematically locate the difference between the sample "3869" and the lower known QC point 5 ppb of "2500".
3869 – 2500 = 1,369 points more than 5 ppb value.
= 2.738 ppb more than 5 ppb = 7.738 ppb

If you use the QC "10 ppb":
10 ppm = 5000
5000 – 3869 = 1,131 graph points
1131 points ÷ 500 (points per 1 ppb) = 2.262 ppb lower than the 10 ppb QC value = 10 – 2.262 = 7.738 ppb

As you see, whether you used the lower or the higher QC, you obtained the same value for the unknown sample trapped between them on a perfect graph.

Now, depending on your sample size (weight or volume) ratio to level of detection, dilution if any, and the accuracy required, and depending on the instrumental sensitivity, a less than a straight line graph, with a C.C. less than .996, even 1% or 2 % difference between the 2 QCs trapping the sample, could mean as little as .001 ppb difference in sample value.

Well, if your expectation is 0.1 ppb then you're Ok and your non-chart mathematical calculation would be as good as a graph.

By the way, this is not something I ever read anywhere and I have no knowledge of any similar written approach. It was something I came up with on the spot one day when an instrument failed to provide a perfect printout. I explained it to fellow assistants who were surprised they never thought of it, and they even continued to refer to it as the "Gouda Graph calculation."

Having mathematical values to be calculated manually is not a case that you would never face. It's a rare concern since the instrument will calculate it, but there will be analytical steps via certain bench methods that would require a graph paper. So, let's simplify it this way:

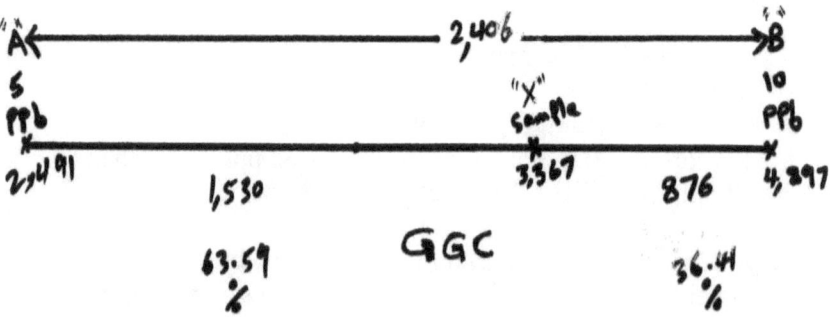

Let's say you have 2 QCs, "A" = 5 ppb & "B" = 10 ppb
5 ppb reads 2,491 and 10 ppb reads 4,897
The area between then is 2,406 graph points representing 5 ppb between the two QC points.

If the sample read 3,367, then:
The sample is 1,530 point more than 5 ppb
And is 876 points less than 10 ppb.

The area between the sample reading and QC-5ppb reading is 63.59% of the total area between the two QCs (the ratio between 1,530 and 2,406.

Similarly, the area between the sample reading and the QC-10ppb is 36.41% of the total area between the two QCs (the ratio between 876 and 2406.)

Meaning, the sample value is 36.41% less than 10 ppb and is 63.59% more than 5 ppb.

Lets look at this example:

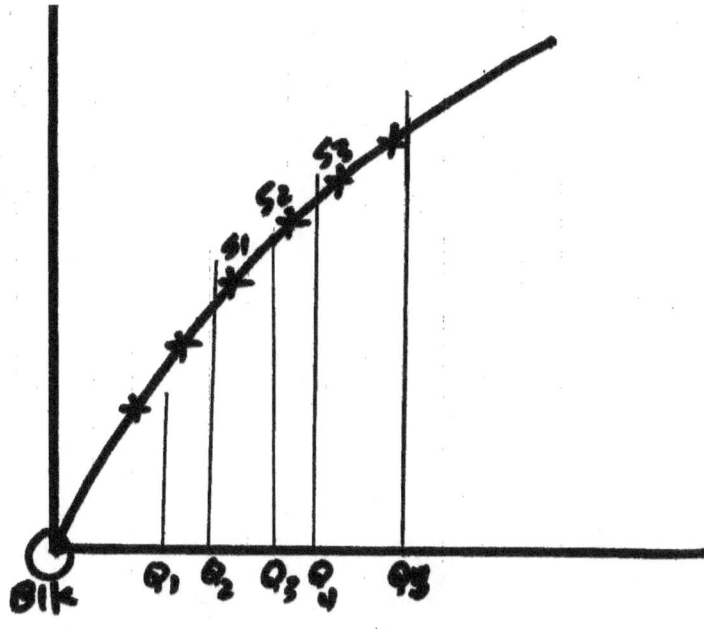

This is clearly less than an ideal graph, however, an overall .990 CC. of the whole graph area, can be manipulated mathematically within a selected area isolated between secured QCs to obtain 0.999 CC. This would allow you to determine – in this case, any of the samples shown, e.g. S1, S2, S3 based on the baseline noise and the blank performance at the bottom (and a low, near DL QC) and the isolated small area of the sample where you can trap a short straight line featuring the sample secured between MM-QCs.

Lab notes:

Biological samples:

The following illustrates the approach to organic samples the chemist will be often faced with.

(0.5 g max. of wet weight)

Sample digestion flask
Distilled water, 1 ml
HNO_3-$HClO_4$ (1+1), 2 ml
H_2SO_4, 5 ml
Heat at 200-230°C for 30 min.
Digested samples
Cool.
Top up to 50 ml with distilled water.
Test solution, a fixed volume (usually 5 ml)
10% $SnCl_2$ solution, 1 ml
CVAAS

Determination of Total Mercury in Biological Samples (fish, shellfish, human hair, blood, and tissues such as umbilical cord)
Sample digestion flask
HNO_3-$HClO_4$ (1+1), 2 ml
H_2SO_4, 5 ml

Urine samples, 2 ml
Add dropwise while swirling slowly.
Heat at 200-230°C for 30 min.
Digested samples
Cool.
Top up to 50 ml with distilled water.
Test solution, a fixed volume (usually 5 ml)
10% $SnCl_2$ solution, 1 ml
CVAAS

Hair:
Weigh 20-30 mg of hair sample in a beaker, wash with neutral detergent (diluted 100-fold) and distilled water by decantation, and wash again with a small amount of acetone to remove the water. Remove the residual acetone under reduced pressure. Transfer the hair sample into a 20 ml vial and cut into an approximately powdery state with dissection scissors to make a sample for analysis.

Reagents

- Acetone: CH_3COCH_3 (analytical grade)
- Ethanol: C_2H_5OH (analytical grade)

- HNO_3-$HClO_4$ (1+1): Mix 100 ml of perchloric acid (for measurement of toxic metals) into 100 ml of nitric acid (for measurement of toxic metals). (Store in a cool dark place.)

- H_2SO_4: Sulfuric acid (for measurement of toxic metals)
- Distilled water: Distill deionized water and store in a clean glass container.

- HCl: Hydrochloric acid (analytical grade)
- 10% $SnCl_2$ solution: Dissolve 10 g of tin (II) chloride dihydrate (analytical grade), $SnCl_2 \cdot 2H_2O$, in 9 ml of HCl and dilute to 100 ml with distilled water. Aerate with N_2 gas (100 ml/min., 20-30 minutes) to expel any mercury from the solution.

- 5N NaOH: Dissolve 20 g of sodium hydroxide (analytical grade) in distilled water to make a final volume of 100 ml.

- 0.1N NaOH: Dilute 5 N NaOH 50-fold with distilled water.
- 0.1% L-cysteine solution: Dissolve 10 mg of L-cysteine hydrochloride, $HSCH_2CH(NH_2)COOH \cdot HCl \cdot H_2O$, in 10 ml of 0.1N NaOH. (Prepare a fresh solution for each analysis.)

- Methylmercury standard solution: Weigh out 12.5 mg of CH_3HgCl (authentic standard) in a 100-ml volumetric flask, dissolve in toluene to make a final volume of 100 ml, and store as stock solution. Dilute the stock solution 100-fold with toluene to obtain a methylmercury standard solution. One ml of this solution contains 1.0 µg of Hg.

- Methylmercury-cysteine solution: Transfer 0.5 ml of the methylmercury standard solution and 5 ml of the 0.1% L-cysteine solution into a 10-ml conical centrifuge tube fitted with a stopper. Shake for 3 minutes with a shaker to extract methylmercury into the aqueous phase. Centrifuge at 1,200 rpm for 3 minutes and draw off and discard the organic phase (upper phase). Seal the tube and store in a cool dark place. (Prepare a fresh solution monthly.) One ml of this solution contains 0.1 µg of Hg.

- 1N H_2SO_4: Gradually add 30 ml of sulfuric acid (for measurement of toxic metals) to distilled water to make a final volume of 1,000 ml.

- 1% acidic $KMnO_4$ solution for collecting mercury: Dissolve 1 g of potassium permanganate (analytical grade) in 100 ml of 1N H_2SO_4.

- 0.5% $KMnO_4$ solution: Dissolve 0.5 g of potassium permanganate (analytical grade) in distilled water to make a final volume of 100 ml.

- Toluene: $C_6H_5CH_3$ (reagent grade for residual pesticide analyses)

Preparation of sample test solution:

Transfer 2 L of a water sample into a 2 L separatory funnel. Add 10 ml of 20N H_2SO_4 and 5 ml of 0.5% $KMnO_4$ solution, mix by shaking.

Let stand for 5 minutes.

Neutralize with 20 ml of 10N NaOH, add 5 ml of 10% NH2OH·HCl solution, and shake.

Let stand for 20 minutes.

Add 5 ml of 10% EDTA solution to the mixture and mix by shaking.

Add precisely 10 ml of purified 0.01% dithizone-toluene followed by vigorous shaking for 1 minute to extract the mercury in the sample.

Let stand for at least 1 hour, avoiding direct sunlight. Discard the aqueous phase (lower phase). Transfer the toluene phase preferably into a 10-ml conical centrifuge tube fitted with a glass stopper and centrifuge at 1,200 rpm for 3 minutes with the glass stopper in place.

When an emulsion is formed, add 0.5 g of anhydrous sodium sulfate and shake followed by centrifugation to remove the lower phase.

Transfer a fixed volume (usually 7 ml) of the toluene phase into a sample digestion flask.

With a rotary evaporator, evaporate to dryness on a water bath at 60°C.

Add 1 ml of distilled water, 2 ml of HNO_3-$HClO_4$ (1+1), and 5 ml of H_2SO_4 and heat on a hot plate at 200-230°C for 30 minutes.

Allow to cool and add distilled water to obtain a fixed volume of 50 ml.

Mix well and use this as a sample test solution. Separately, choose a sample with a lower mercury content based on the type of water sample.

To each of the 2-L water samples chosen, add 0 and 0.2 ml of methylmercury-cysteine solution (100 ng Hg/ml; corresponding to 0 and 20 ng Hg).

Follow the above preparation procedures for sample test solutions to obtain a blank test solution and standard test solution for the measurement of total mercury.

Test procedures and calculation

Transfer a fixed volume, V ml (normally 10 ml) of the blank test solution, standard test solution, and the sample test solution into the reaction vessel in the mercury analyzer and apply the stopper.

Add 1 ml of 10% $SnCl_2$ solution with the accessory dispenser and push the start button.
The diaphragm pump will run and the generated elemental mercury vapor will be circulated through the 4-way cock between the reaction vessel and the acidic gas trap for 30 seconds to homogenize the concentration in the gas phase, while the acidic gas generated from the test solution is collected in the alkaline solution.

After 30 seconds, the 4-way cock will turn automatically by 90°, allowing the introduction of the elemental mercury vapor into the photo-absorption cell through the ice bath for measurement of the absorbance.
The readings of the recorder will increase sharply and decrease with a sharp peak. When the reading of the recorder begins to decrease, open the cock on the lower part of the reaction vessel to discard the solution inside, close it again, and allow it to aerate until it returns to the baseline. Push the reset button to start the next measurement

Notes:

The peak heights (mm) obtained after measurement of fixed volumes V ml (normally 10 ml) of each of the blank, the standard, and the sample test solutions (or their diluted solutions) are labeled Pbl, Pstd, and Ps, respectively. The total mercury concentration in the sample is calculated with the following formula:

Total mercury concentration in the sample (ng/L) =

20 ng × (Ps–Pbl)/(Pstd–Pbl) × dilution factor × 1/ sample volume (L)

Procedural Notes

1. Since concentrations of mercury in water samples are extremely low and usually at the ng/L level, pre-concentration of the mercury in the sample is required for measurement of the mercury. In the present method, quantitative and efficient pre-concentration is performed by extracting the mercury with a small volume of 0.01% dithizone-toluene after ionization of all the mercury species in the sample with potassium permanganate in an acidic medium with sulfuric acid.

This uses the chemical properties of dithizone: it combines easily with ionic mercury species to form a complex salt that is insoluble in water but soluble in organic solvents such as toluene. After pre-concentration by extracting the mercury with 0.01% dithizone-toluene, the extract is evaporated to dryness under reduced pressure. Similar to the case of biological samples, the residue is subjected to wet digestion to prepare a sample test solution for the total mercury analysis with CVAAS.

2. Dithizone (diphenylthiocarbazone) is easily oxidized and usually contains its oxidized form, diphenylthiocarbadiazone, as an impurity. As well, it sometimes contains mercury or the like in the form of metal complex, although the amount is minute.
These impurities are highly soluble in organic solvents but insoluble in alkaline solutions, whereas pure dithizone has the chemical property of being soluble not only in organic solvents but also soluble in alkaline solutions by forming its salts. This property enables impurities to be removed and dithizone to be purified for use.

3. For samples containing large amounts of Cl^- ions such as seawater, the treatment of the sample with a combination of potassium permanganate and sulfuric acid causes oxidation of Cl^- ions to Cl_2 to occur during the treatment, and the resulting

Cl_2, once generated, is difficult to reduce by treatment with hydroxylamine hydrochloride solution.

This results in oxidation of dithizone in the subsequent dithizone-toluene extraction step. Therefore, particularly for seawater samples, it is important to maintain the 5 minutes treatment time with potassium permanganate. Further, after the addition and mixing of the hydroxylamine hydrochloride solution, allow at least 20 minutes of reaction time before the EDTA treatment and dithizone-toluene extraction procedures.

Sample, 2 L (2 L separatory funnel)
Add 10 ml of 20N H_2SO_4 and mix.
Add 5 ml of 0.5% $KMnO_4$ solution and mix.
Let stand for 5 min.
Add 20 ml of 10N NaOH and mix to neutralize.
Add 5 ml of 10% $NH_2OH \cdot HCl$ solution, mix, and allow to stand for 20 min.
Add 5 ml of 10% EDTA solution and mix.
Add 10 ml of purified 0.01% dithizone-toluene and vigorously shake for 1 min.
Allow to stand for at least 1 hr.
Organic phase(10-ml conical centrifuge tube) Aqueous phase (When an emulsion is formed, add 0.5 g of Na_2SO_4 and shake.)
Centrifuge at 1,200 rpm for 3 min.
Organic phase, 7 ml (sample digestion flask)
Evaporate to dryness.
Residue
Distilled water, 1 ml
HNO_3-$HClO_4$ (1+1), 2 ml
H_2SO_4, 5 ml
Heat at 200-230°C for 30 min.
Digested sample
Allow to cool.
Top up to 50 ml with distilled water.
Test solution, a fixed volume (usually 10 ml)
10% $SnCl_2$ solution, 1 ml
CVAAS

Notes

Analytical Method for Methylmercury

For measurement of organic mercury, gas-liquid chromatography with electron capture detection (GLC-ECD) is used for selective analysis of methylmercury and other organomercury compounds. Because this technique provides good separation and superior sensitivity for analyzing organomercury halides, it has been widely used for the determination of methylmercury in various kinds of biological and environmental samples.

The analytical procedure involves the extraction of methylmercury in the samples as its halide into an organic solvent after acidification; the back-extraction into a cysteine- or glutathione-aqueous solution; the re-extraction into an organic solvent; and measurement of methylmercury by GLC-ECD.

As an alternative, methylmercury can be determined by CVAAS, which measures elemental mercury vapor generated from a heated sample test solution obtained from similar methylmercury extraction procedures. However, in this direct extraction procedure with organic solvent, a solid emulsion is often formed during the extraction process, particularly with fish and other biological samples.

This makes the following steps complicated and causes the extraction efficiency of methylmercury to vary with the type of sample. While several pre-treatment methods are proposed to overcome the above drawbacks, we describe herein the following two methods: determination by the dithizone extraction/GLC-ECD method, which is suitable for methylmercury in various types of biological and environmental specimens; and the hydrochloric acid leaching/toluene extraction/GLC-ECD method for the determination of methylmercury in hair.

Determination by the dithizone extraction/gas-liquid chromatography with electron capture detection (GLC-ECD) method.

Principle:

The dithizone extraction/GLC-ECD method was established as an analytical method for methylmercury in various biological and environmental matrices. It is based on the following two advantages of dithizone extraction, which was widely used for colorimetry of inorganic and organic mercury species prior to the introduction of atomic absorption in the late 1960s: it has much higher extraction efficiency than that of direct solvent extraction, facilitating the extraction separations of trace amounts of mercury from samples with a small portion of the solution; and alkyl mercuric dithizonate, as soon as it is injected into GLC, reacts with Cl- in the column to give its chloride form for quantitative detection.

This method involves the pre-treatment of the sample, dithizone extraction, back-extraction into alkaline sodium sulfide, dithizone re-extraction, and GLC measurement.

Appropriate pre-treatment to accommodate the characteristics of the composition of each sample enables efficient extraction of methylmercury with a small portion of dithizone-toluene solution.

After dithizone-toluene extraction, test solutions are prepared with all common procedures, followed by measurement with GLC-ECD. To accommodate the principle used for this method, pack several centimeters of sodium chloride at the injection port of the column on top of the packing material for GLC.

Notes:

- 1N KOH-ethanol: Dissolve 56.11 g of potassium hydroxide (analytical grade) in ethanol to obtain a final volume of 1,000 ml. (Store in a cool dark place.)

- 1N HCl: Mix 90 ml of hydrochloric acid (analytical grade) with distilled water to obtain a final volume of 1,000 ml.

- 20% $NH_2OH \cdot HCl$ solution: Dissolve 20 g of hydroxylamine hydrochloride in distilled water to make a final volume of 100 ml.

- 20% EDTA solution: Dissolve 20 g of tetrasodium ethylenediaminetetraacetate (analytical grade), $C_{10}H_{12}N_2O_8Na_4 \cdot 4H_2O$, in distilled water to make a final volume of 100 ml.

- 1N NaOH: Dissolve 40 g of sodium hydroxide (analytical grade) in distilled water to make a final volume of 1,000 ml.

- 0.1N NaOH: Dilute 1N NaOH 10-fold with distilled water.

- Purified 0.01% dithizone-toluene: Dissolve 0.011 g of diphenylthiocarbazone, $C_6H_5N:NCSNHNHC_6H_5$, in 100 ml of toluene in a 200-ml separatory funnel. Add 50 ml of 0.1N NaOH and shake briefly to extract the dithizone into the aqueous phase (bottom phase). After allowing the phases to separate, transfer the bottom phase into a glass container fitted with a glass stopper.
Add 1N HCl dropwise to make the solution slightly acidic (blackish-green crystals will precipitate). Add 100 ml of toluene and shake to obtain purified 0.01% dithizone-toluene. Allow the phases to settle, draw off and discard the bottom phase, and seal. Store in a cool dark place. (Prepare a fresh solution for each analysis.)

- Alkaline sodium sulfide solution: Weigh out 0.15 g of $Na_2S \cdot 9H_2O$ (analytical grade) in a 10-ml conical centrifuge tube with a glass stopper and dissolve in 10 ml of distilled water to make the sodium sulfide stock solution. (Prepare a fresh solution monthly. Store in a cool dark place.) At each use, transfer 0.1 ml of the stock solution into a glass container with a glass stopper, add 50 ml of 0.1N NaOH and 50 ml of ethanol, and mix to make an alkaline sodium sulfide solution. (One ml of this solution contains 5 µg of Na_2S.)

- Walpole's buffer: Mix 200 ml of 1M CH_3COONa and about 200 ml of 1N HCl in 600 ml of distilled water to adjust to pH 3.0.

- Anhydrous sodium sulfate: Anhydrous sodium sulfate (pesticide analysis grade) heated at 500°C for 2-3 hours (stored in a desiccator)
- 0.1% L-cysteine solution: Dissolve 10 mg of L-cysteine hydrochloride, $HSCH_2CH(NH_2)COOH \cdot HCl \cdot H_2O$, in 10 ml of 0.1N NaOH. (Prepare a fresh solution for each analysis.)

- Methylmercury standard solution: Weigh out 12.5 mg of methylmercury chloride, CH_3HgCl (authentic standard) in a 100-ml volumetric flask, dissolve in toluene to make a final volume of 100 ml, and store as a stock solution. Dilute the stock solution 100-fold with toluene to make the methylmercury standard solution. One ml of this solution contains 1,000 ng of Hg.

- Methylmercury-cysteine solution: Transfer 0.5 ml of the methylmercury standard solution and 5 ml of the 0.1% L-cysteine solution into a 10-ml conical centrifuge tube with a stopper. Shake for 3 minutes with a reciprocal shaker to extract the methylmercury to the aqueous phase. Centrifuge at 1,200 rpm for 3 minutes with the glass stopper attached and draw off and discard the organic phase (upper phase). Seal the tube and store in a cool dark place. (Prepare a fresh solution monthly). One ml of this solution contains 0.10 µg of Hg.

- Florisil: Florisil for column chromatography (60-100 mesh) heated at 130°C for 2-3 hours (stored in a desiccator)

- Florisil column: A glass column packed with 0.5 g of Florisil (60-100 mesh) and 0.5 g anhydrous sodium sulfate in turn.

- 2N HCl: Mix 180 ml of hydrochloric acid (analytical grade) with distilled water to obtain a final volume of 1,000 ml.

- 1N NaOH: Dissolve 40 g of sodium hydroxide (analytical grade) in distilled water to make a final volume of 1,000 ml.
- 0.1N NaOH: Dilute 1N NaOH 10-fold with distilled water.

- 0.1% L-cysteine solution: Dissolve 10 mg of L-cysteine hydrochloride, $HSCH_2CH(NH_2)COOH \cdot HCl \cdot H_2O$, in 10 ml of 0.1N NaOH. (Prepare a fresh solution for each analysis.)

- Methylmercury standard solution: Weigh out 12.5 mg of methylmercury chloride, CH_3HgCl (authentic standard) in a 100-ml volumetric flask, dissolve in toluene to make a final volume of 100 ml, and store as stock solution. Dilute the stock solution 100-fold with toluene to make the methylmercury standard solution. One ml of this solution contains 1,000 ng of Hg.

- Methylmercury-cysteine solution: Transfer 2 ml of the methylmercury standard solution and 2 ml of the 0.1% L-cysteine solution into a 10-ml conical centrifuge tube with a glass stopper. Shake for 3 minutes with a reciprocal shaker to extract the methylmercury into the aqueous phase. Centrifuge at 1,200 rpm for 3 minutes with the glass stopper attached and draw off and discard the organic phase (upper phase). Seal the tube and store in a cool dark place. (Prepare a fresh solution monthly). One ml of this solution contains 1,000 ng of Hg.

Notes:

Back to fundamentals:

Hydrargyrum compounds show both +1 mercurous & +2 mercuric oxidation states.

The mercurous compounds contain two Hg atoms bound together. In aqueous solutions, the ion is double ion corresponding to Hg^{2+}_2 in which there is a covalent Q bond between the two atoms.

Experimental evidence for this is the lack of paramangnetism of mercurous compounds. The ion Hg+ would have one unpaired electron in its 6s orbital and would be paramagnetic, whereas ion Hg^{2+}_2 would have the two electrons paired in a Q bonding molecular orbital and would form an equilibrium between liquid mercury, mercuric ion, and mecurous ion as follows:

$Hg + Hg^{2+} = Hg^{2+}_2$

$Hg + Hg^2 = 2Hg^+$

Mercurous ion behaves much like Ag+, e.g. it reacts with chloride ion to precipitate white mercurous chloride, Hg_2CL_2 (calomel). When exposed to light, calomel darkens by partial disproportional into Hg and $HgCL_2$. Just as silver halides decrease in solubility in going from AgF to AgI, so do mercurous halides.

Mercurous fluoride, Hg_2F_2 is quite soluble in H_2O, but the solution quickly decomposes to form HF and insoluble black Hg_2O. For the other halides the solubility products are as follows:

$$Hg_2Cl_2 = Hg^{2+}_2 + 2\,Cl^-$$

$$Hg_2Br_2 = Hg^{2+}_2 + 2Br^-$$

$$Hg_2I_2 = Hg^{2+}_2 + 2\,I^-$$

Unlike Ag^+, mercurous ion does not form an ammonia complex. When aqueous ammonia is added to Hg_2Cl_2 the solid turns black because of the formation of finely divided mercury :

$$Hg_2Cl_2 + NH_3$$
$$= HgNH_2Cl + Hg + NH^+_4 + Cl^-$$

When performing test-tube bench chemistry, note that the compound **$HgNH_2Cl_2$** mercuric ammonobasic chloride, is white, yet its colour is obstructed by the intense black of the mercury.

In the +2 state, mercury is frequently represented as the simple ion Hg^{2+} although it is usually in the form of complex ions, insoluble solids or weak salts. For example, in a solution of the weak salt mercuric chloride, the concentration of Hg^{2+} is much smaller than the concentration of non-dissociated $HgCl_2$

molecules. With excess chloride ion the complex $HgCl_3^-$ and $HgCl_4^{2-}$ are also formed.

Well ... having said all that, let us get back to the cold vapour technique.

Notes

Back to cold vapour
Analytical notes:

As stated earlier, the "cold vapor" technique is based on the fact that mercury is reduced to the elemental state. Elemental mercury vapor is aerated from solution in a closed system and absorption (peak height or peak area) is measured as the function of mercury concentration.

The following technical notes are pertinent facts & factors:

* Hg atomic absorption at 253.7 nm wavelength constitutes no *complex spectra* concern.
Specific interferences are indicated in this paper.

* Standard solutions / Q. C. references need be prepared from different sources for cross-
check purposes. Typical references are 1000 ppm solutions from 2 sources {such as
BDH, Plazmachem and Optimum Green.} and at least one source - typically a salt, e.g.

$HgCL_2$, $Hg(NPO_3)_2$ or CH_3HgCL .

* Aqueous samples are first acidified to pH < 2 with HNO_3. Non-aqueous samples are
refrigerated. **$K_2Cr_2O_7$** is used as a preservative when preparing stock standard
solutions.

It is of course important that samples and QC standards undergo the same treatment; e.g. the same $K_2Cr_2O_7$ solution, out of the same bottle would be used in the preparation of the blank standard and Hg QC samples as well as additional spiked samples. In the attached analysis illustration, chemist-author Paul Gouda named the zero-blank: Q1, calibration standard closest to sample concentration (or two standards, a&b below and above sample concentration): Q2, spiked sample: Q3 (including spiked blank, applicable to both sets of samples; strict standards and matrix match standards.

* Organic mercury compounds will not respond to cold vapor technique unless first broken down and converted to mercuric ions vulnerable to reduction to ground state atoms by $SnCL_2$. $KMnO_4$ acts as a strong oxidant to serve this purpose.

Some organic mercurial *such as phenyl mercuric acetate and methyl mercuric chloride* **may require a second treatment with potassium persulphate (as an additional oxidant) added after the regular treatment with $KMnO_4$ in order to complete the oxidation.**

Potassium permanganate will also remove most of the aromatic and nitrogen compounds that could interfere with the photo-metric measurement.

* The method known here in North America as EPA #245 uses sample aliquots weighed into digestion test tubes. The preparative digestion varies based on sample matrix/nature. The procedure provides a sound guideline, again, with the understanding that individual samples may require **individual attention.** Usually a sample of 25ml or 0.5g is treated with 2ml of HNO_3 and 5ml of H_2SO_4, followed by 5ml of 5% $KMnO_4$ solution.

Depending on the sample, an additional volume of an oxidising agent (e.g. HNO_3) may be required. In some cases up to 20ml of $KMnO_4$ had to be added to ensure complete oxidation (purple colour persists). The sample is often digested in a water bath (at 80 degree Celsius) for 90 minutes.
* Special attention to the temperature and the method of heating is necessary to prevent **volatile mercury in elemental state** from escaping. Some organic samples (where Hg is in a compound form, not in elementary state) require heat block digestion at a higher temperature and over a longer time in order to break down the organic compound. Clearly high temperature digestion is ideal for some samples while low temperature digestion is a must for others as I'll explain latter.

Preparative digestion in biological samples is designed for degradation of samples to free mercury from chemical bonds and biological matrices. Fuming with H_2SO_4 at a temperature as high as 260 °C produced 97 % recovery. Without such fuming the organic compounds did not breakdown and recovery was < 45 %.

At the same time, samples with volatile mercury in elementary state need close temperature control. One

must keep in mind that the solid method is based on dealing with mercury in Hg II state.

Other methods use a *perchloric digestion* at 200 - 260 C . This is reported to succeed with organic Hg compounds such as methylmercury. It has also been reported that **$HCLO_4 + HNO_3$ digestion** followed with **H_2O_2** treatment recorded good recovery. I have experimented with *aqua regia* digestion and obtained very good recovery; which I'll address in details in part-II of this study "next publication".

* Digestion is carried out in 50ml test-tubes. Samples are allowed to cool to room temperature and 5ml of 12% hydroxlamine solution (12% w/v sodium chloride & 12% w/v hydroxlammonium chloride) is added to reduce excess of permaganate {reddish colour clears}.

In some cases up to 15ml of $NaCL.HO-NH_3CL$ had to be added. Individual attention in terms of digestion approach, i.e. volume of and specific oxidising and reducing agents utilized, temperature and length of digestion, is a crucial judgement call and a vital factor in achieving successful oxidation and reduction and thus a good result.

* When working on highly organic samples [e.g. fish and lobster] it is necessary to digest the sample with HNO_3 [conc./neat] on hot block for a longer period of time until **NOx** are expelled (brown fumes cleared). An additional volume of $KMnO_4$ is a precaution against possible mercury dissipation.

This will be especially necessary for samples containing high fat and cellulose because lipids / complex compounds are not easily broken down especially at a lower temperature.

It should also be noted that with some samples, the permanganate solution used in digestion should be allowed to sit at room temperature [exposed to light] for some two days to allow MnO_4 that may form to settle out.

Generally, a saturated permaganate solution (or 5% w/v) is used as a second oxidant especially to complete the oxidation of organic material. An additional volume of $KMnO_4$ is often needed during digestion when there is a possibility of formation of **nitrogen compounds** that could interfere with the photo measurement by AAS. This is only one of many examples of the importance of the often necessary individual treatment to certain samples.

* The hydroxylamine solution is used mainly to reduce excess permanganate.

Experiments indicate that a concentration of 12% w/v concentration [12% sodium chloride and 12% hydroxylam - ammonium chloride]. **NaCL.HO-NH$_3$CL** is a sound guide-line.

Notes

Major reagents:

HCL: neat {conc.}

$SnCL_2$: 5 % w/v depending on sample as explained below.
The common Tin [II] chloride concentration is 10 % $SnCL_2.2H_2O$ in 20 % HCL.

H_2O: distilled and deionized.

($NaCL+NH_2OH.HCL$): 12% w/v

$KMnO_4$: 5%

H_2SO_4: neat-conc.

Argon gas is the medium by which Hg atomic vapour is swept into the cell [using a peristaltic pump producing a rate of 1L/min.] Nitrogen can be used as an insert gas. In addition, compressed air can be used as a substitute carrying agent, however it has to be treated to ensure being moisture free [e.g. filtration through magnesium perchlorate]. Special attention should be given to the cell to ensure being dry and clean and thus avoid spectrophotometric interference.

Condensation of water vapour in the cell can cause attenuation of the light beam. This problem can be circumvented with proper handling technique of the vapour generator including allowing for the final post-delivery gas purge which, in this stage, is not impregnated with moisture. Also proper maintenance, pump calibration [reagent control & reaction time], argon pressure, setting of the generator unit...etc. are all contributing factors to successful analysis.

The pump can be easily calibrated using 100ml of H_2O in the empty reducing agent reservoir. The Varian generator accessory used in this experiment is calibrated by placing a funnel over it, and place the sample fleaker cap over the funnel. Press start/reset button and hold the head over the funnel until the circle is complete and the pump ceases to operate.

The volume of H_2O in the graduated cylinder should be approximately 5ml. Special attention should be given to securing the beaker cap to prevent mercury vapour from escaping.

* It is important to prepare fresh reducing agent [Stannous Chloride] to ensure stability of $SnCL_2$. The agent is normally prepared in 5 M HCL. Practically 5 % v/v of Hydrochloric acid is sufficient to keep the Stannous Chloride in solution, however it should be noted that an *increase of $SnCL_2$ concentration contributes to increasing the absorption sensitivity.*

Experiments indicate that raising the Stannous Chloride up to 20% w/v gives the highest signal. No improvement in sensitivity was detected beyond 25% $SnCL_2$ concentration.

25% $SnCL_2$ solution can be easily kept in 20% HCL. Most of the time a 10 % $SnCL_2$ in 10% HCL is sufficient. There are however samples that require a higher concentration of $SnCL_2$ solution when a higher signal / better sensitivity is needed even at the cost of a slightly higher background noise and / or a less stable curve.

The analyst has to consider his options and balance his approach accordingly. It is one of those judgement calls by the chemist who has to consider the factors involved and the special application of the sample. It must however be noted that consistent calibration is imperative. EPA method #245 recommends 20% $SnCL_2$ w/v in 25% HCL v/v for reason of solubility and stability.

H_2O condensation is reported as a major spectra interference. This necessitated ensuring the use of dry compressed air via **dehydration with granular calcium sulphate trap ($CaSO_4$) or $Mg(CLO_4)_2$.** The use of argon to substitute the compressed air appears to be a safer choice. As I indicated, nitrogen can also be used as a purge gas. A 30 psi pressure setting is reported to be ideal.

* The extreme sensitivity of the analytical
   procedure and the omnipresence of Hg
necessitate extreme care to avoid extraneous contamination. Glassware need to be cleaned in HNO_3 for 30 minutes. Deionized - distilled H_2O {ideally 3 times} is used, and when necessary soaking in a 20% w/v sodium EDTA for 1 hour would be a good approach .

Methyl mercuric chloride is partially retained on the sample and is, therefore a potential interference. A number of organo - mercury compounds including phenyl mercuric acetate and methylmercuric chloride are only partially oxidized by potassium permanganate.

Experiments show that using potassium persulfate {in addition to and following $KMnO_4$} insures that such organic Hg compounds are oxidized to mercuric ions.

This is the recommendation of the EPA method which claims that the addition of the persulfate as a second oxidant increased recovery up to 98% when treating difficult organic samples.

Personal experience indicates that this is only true with very specific sample matrix (this issue will addressed in paper II), otherwise, the difference is very insignificant when treating inorganic samples with "permanganate only" as opposed to addition of both "permanganate and persulfate" - provided that the chemist has done everything right; and I'll elaborate further in paper II.

* Sea water, brines and industrial effluents high in chloride require additional permanganate {up to 25 ml}. *During the oxidation, chlorides are converted to **free chloride** which will absorb radiation at 253 nm.* **Care must be taken to assure that free chlorine is absent before mercury is reduced and swept into atomic absorption cell.** This can be accomplished by using an excess of hydroxylamine solution {up to 25 ml}.

Possible **interference from sulphide** is eliminated by KMnO$_4$. Experiments show that concentration as high as 20 ppm of sulphide (sodium sulphide} did not interfere with the recovery of added inorganic mercury using spiked distilled water samples.

Cu interferes with Hg recovery, however, experiments show that Cu as high as 10 ppm has no effect whatsoever on the recovery of spiked samples. No Cu interference was detected until its concentration at the % level.

Interference caused by certain *volatile organic materials which will absorb radiation at this wavelength* has been reported. This necessitates a preliminary matrix-match background run {analytical blank} to determine if such interference is present {false positives}. I'll address the concerns related to suppressants {false negatives} in a future article. Meanwhile, matrix match blank will aim at the false positive concern, and the standard addition approach would be the obvious recommendation to determine if false negative is a problem.

It must also be noted that reduction techniques that produce a transient - i.e. short-lived absorbance signal, create a need for a suitable auto integration time set up {and a fast-response recorder, of course}.

This becomes a major factor when comparing two different cold vapour approaches such as method ESS #

H10 referred to in last fall's issue of "The European scientist" magazine as opposed to the method referred to here in as adopted by EPA.

When organic material interfere, it is advisable to analyze the sample both by using the **regular procedure and again under oxidizing conditions only** - that is without the reducing reagents. The true Hg value would be the difference between the two values.

Typical interference were found to exist within this guideline:

Ni^{+2} : 20% enhancement

Cr^{+6} : 200 ppm

Fe^{+3} : none // 1000 ppm

Cu^{+2} : none / 1000 ppm

KI : 1 ppm

HNO_3 : none / 20 %

H_2SO_4 : none / 20 %

Hydrargyrum analysis – Dr. Paul Gouda

Other interference:

1) H_2O vapour condensation.

2) Chlorides and oxides of nitrogen.

3) Some diatomic gases such as CL_2.

4) Volatile aromatics such as acetone and chloroform

5) Certain elements are also reported to interfere chemically causing a low bias, e.g. Se, Te , Sb, Bi and As.

HCL is the orthodox choice of acid. "Varian" experiments show that 5% HCL + 5% HNO_3 matrix gave the best sensitivity while 5% HNO_3 gave the poorest signal.

"The British report" indicates that dichromate used for stabalization of mercury standard solutions causes a drop of 7% in sensitivity at the concentration level of 0.1% and this is corrected by preparing Hg standard solution at higher concentration level (to necessitate a higher dilution for working standard solution, i.e. lower $K_2Cr_2O_7$, and, if necessary, increasing the HNO_3 in the stock standard solution to a much higher level. The European method recommends a 40% HNO_3 for stock Hg standard solutions.

* A major control sample is the spike Q.C. for obvious reasons. Spike recovery will give a good indication whether the sample matrix is acting as a suppressant. The sample is spiked with a known value of mercury and the ideal recovery would of course be the total of original sample concentration plus the injected additional spike value.

Resorting to comparing the absorbance difference between sample slope and correction curve corrects the problem. The correct graph should consist of at least 3 different spiked values.

In addition, the graph must present the difference between the two blanks {**analytical blank and arbitrary blank**}. This will ensure producing a confirmed, stable actual calibration curve with > 95% correction factor.

The sensitivity of the method is also affected by other parameters such as **temperature** of the reducing solution and the flow of the carrying gas. An increase of the temperature of the solution produces more sensitivity. As indicated, concentration of the Tin II chloride solution and concentration of HCL contribute to signal sensitivity.

* The majority of sample preparations, at the AAS stage, are likely to contain trace level Hg < 0.5 ppb in solution. Increasing Hg level too high in the prepared sample by taking a large sample, in most cases, will introduce larger background factor. The practical detection limit of the method is said to be 0.2 ppb, hence, readings below 0.2 ppb can be technically dismissed and reported as **BDL** {below detection limit}.

However, with proper sample digestion and instrument operation, the chemist can easily determine an obvious 0.1 ppb peak with obvious certainty; and; I personally, so did fellow chemist at 2 laboratories, have repeatedly, consistently and positively detected a 0.05 ppb presence with satisfactory assurance.

My Q.C. references have always included a 0.05, 0.1, 0.25 and 0.5 ppb standards all of which have given obvious confirmed reference peaks. With the addition to the reference "0" sample {blank} and spiked samples at the corresponding level of concentration, the chemist can easily determine trace level 0.05 - 0.2 ppb with great certainty.

* A typical calibration curve is 0 "blank", .5, 1, 2, 3, & 5 ppb standards with **a minimum correlation coefficient factor of 0.995**.

The base line must be monitored and any shift must be corrected. It must also be determined whether a deviation at base line is a calibration shift or a fluctuation and thus data must be edited accordingly after a confirmatory test using known standards is established.

* Experiments show that other reducing agents such as hydrazine hydrate 40% or sodium borohydride 5% do not produce as good recovery as Tin II chloride {$SnCl_2$ 10% in 10% HCL}:

$$Sn^{2+} + 2HgCL_2 = Hg_2CL_2 + Sn^{4+} + 2\,CL^-$$

$$Sn^{2+} + HgCL_2 = Hg + Sn^{4+} + 2CL^-$$

AAS setting:
A basic illustrative AAS set up for high level Hg.

pattern: gain; 550 / cathode lamp: 5 / bandwidth: 0.3

The bench chemist must keep in mind that the baseline absorbance noise is related to the light source intensity. The baseline noise decreases as the source intensity increases. Intensity of light source Io is proportional to the square of lamp current. Accordingly, when lamp current is increased, the baseline noise level decreases. However, if the lamp current is over increased, the phenomenon known as self absorbance becomes a concern.

In one case, while using an old lamp, spectral interference was obvious with a couple of samples.
The standard addition approach (blank spiked with known low QC value) confirmed that there is a problem. The false reading was a result of a different atomic absorbance that fell within the width of the monitored element's absorbance line. This, of course, resulted in a "false positive" and one would have to review that sample digestion and preparation process form step one to the atomic absorbance stage.

The analyst must also keep in mind that the cathode lamp can be affected by metal impurity in the cathode itself (or infected from the "W" anode). This is a common case that happen with old lamps. Narrowing the slit width would correct the problem and eliminate the false positive. A spiked Q.C. sample should then be repeatedly used as confirmatory test.

Another common problem that was experienced with a deteriorating lamp was its inability to produce sufficient voltage impressed across the electrodes, i.e. its **inability to ionize sufficient Ar atoms enough to bombard the cathode and thus produce an effective electromagnetic radiation (energy beam)**. This necessitates alterations to the AA setting to compensate for the problem. Of course, this shouldn't happen - as the analyst must use a healthy cathode lamp.

Of course, when making such adjustments, the analyst again must keep in mind that an increase in hollow cathode current results in an increase in the kinetic energy of the ionized fill gas "Ar" causing more atoms to be sputtered.

As the population of the sputtered atoms increases, the residual unexcited atoms cool and a cloud of neutral atoms in front the cathode is formed. These neutral atoms absorb some of the lamp light which results in an attenuation of the resonance radiation resulting in a classic case of self-reversal or self-absorption.

It should also be noted that spectral band width is also a major factor in signal-to-noise ratio. **A large spectral width will generate an excellent ratio, however, the resonance line may not be isolated from other lines and as a result, the calibration curve will not be as linear. At the same time, the good resolution of a "too narrow" spectral band width will not compensate for the poor signal-to-noise ratio due to the considerate reduction of light.**

With certain samples {trace level recovery when it is necessary to determine near BDL concentration} the spectral band has to be opened wider in order to lower the background noise.

In such a case, the sample has to be re-examined under a narrower bandwidth to make certain that the wider monochromater slit did not admit **non-absorbable radiation defraying linear range.**

Voltage / gain for a CV-AAS must be set to compromise allowing the photomultipllier to give sufficiently high current output without adding excessive noise. That is not to say
that it must be always set blindly in the centre of the traditional (green zone).. nor does it mean that it should always be set on - say "620" just because the analysis involves the same parameter! A judgement call is often needed based on the needs and nature of a specific matrix. At times, raising the actual signal peak at the expense of raising the baseline noise is a good choice, while a lower peak that minimizes background noise may in another case be the right choice.

Mercury salts

Among many inorganic and organic mercury compounds, a few are commonly precipitated or interfere during analysis of the "odd" sample. The following notes will help the analyst spot such salt and approach his sample accordingly:

- Mercuric acetate: **$Hg(C_2H_3O_2)_2$**. White water-soluble crystalline powder.

- Mercuric carbonate (basic): **$HgCO_3.3HGO$**. Brown precipitate.

- Mercurous chloride: **Hg_2CL_2**. White powder insoluble in water.

-   Mercuric chloride: **$HgCL_2$**. Very soluble in water or methyl alcohol or ethyl alcohol.

-   Mercurous iodide: **Hg_2I_2**. Bright yellow amorphous powder.

-   Mercuric iodide: **HgI_2**. Bright red tetragonal powder dissolves in alkalis to form complex salts: **$NaHgI_4$ or K_2HgI_4**.

- Potassium iodomecurate dihydrate: **K$_2$(HgI$_4$).2H$_2$O**. Yellow water-soluble.

- Cuprous iodomercurate: **Cu$_2$(HgI$_4$)**. Bright red, water-insoluble.

- Mercurous nitrate: **Hg$_2$(NO$_3$)$_2$**. White monoclinic crystalline, soluble in cold dilute
    nitric acid.

- Mercuric nitrate: **Hg(NO$_3$)$_2$**. Colourless deliquescent crystalline.

- Mercurous sulphate: **Hg$_2$SO$_4$**. Pale yellow.

- Mercuric sulphate: **HgSO$_4$**. Colourless compound soluble in acidic solutions, but decomposes by water to form yellow, water-insoluble basic sulfate **HgSO$_4$.2HgO**.

- Mercuric sulfide: **HgS**. Exists in two stable forms, the black cubic tetrahedral form (soluble mercuric salts + sulfides) and red hexagonal form (cinnabar/vermilion as
found in nature).

In conclusion, I'll present a few selected facts and factors that constitute problems in areas that I have often been consulted with by laboratories and refinery operations.

High temperature digestion without careful preparation is risky and should be limited to such samples as fish or with high fat content. Chemist Jim Bishop, in his paper ***"high temperature acid digestion of mercury in environmental samples - 1975"*** referred to researches conducted by the Canadian Fisheries Research as well as by EPA.

The researches conclude the following:

- High temperature digestion - hot plate / 260 ºC / 4:1 sulphuric & nitric- gave better recovery than water bath digestion .

- The experiments included the following mercury compounds:

   Methyl mercury chloride

   Methyl mercury bromide

   Phenyl mercuric acetate

   Mercuric chloride

   Mercuric sulfate

   Mercuric nitrate

   Mercuric oxide

The experiments also involved oily fish samples such as sturgeon (Acipenser fulvescens) and crap (Cyprinus carpio). The average result was 99.6 % recovery at 5.7 % RSD with **no measurable loses of mercury due to volatilization** despite the high temperature of the aluminium hot plate digestion.

The paper argues that high temperature digestion is certainly more successful than the low temperature (80 ^oC) water bath treatment. Many laboratories have a misconception about how and when mercury is volatile; properly not truly understanding the difference between being in an elemental state versus an organo compound!

Several universities in Canada, Ireland, England, USA, Austria, Egypt and Spain published lengthy reports on their in-house research. Several approaches were experimented with, including:

- Powerful oxidants such as vanadium pentoxide

- Potassium permanganate + sulfuric acid {and hydroxylamine hydrochloride}

- Sulfuric + nitric + sodium molybdate

- 1:1 nitric + perchloric

- Nitric + hydrogen peroxide {260 0C / potassium permanganate / hydroxlamine}

- Aqua regia .....

As well, another issue that must be addressed is the fact that there are several other organo mercury compounds that may take active part in routine test-tube digestion including **phenylmercuric acetatate, chloromethoxypropylmercuric acetate and alkyl mercuric compounds.**

However, having said that, the practical need to worry about such treatment when dealing with common mercury samples during test-tube digestion and analysis is rare; and, chances are, a good beaker or a test-tube digestion - especially with the precautions I addressed under cold vapour - atomic absorption, when applied correctly to the nature of the individual sample by a keen chemist, would eliminate such interference factors. QC records prove this claim to be true.

This book however is meant to provide a complete reference covering all the rare (yet real) possible complications that have been experienced or reported by or to the author.

I intend to write a comparative study addressing the use of the preceding oxidants and analyzing each approach's advantages and disadvantages.

In conclusion, I can simply state that cold vapour technique, when applied correctly, is a reliable atomic absorption choice for ultra trace analysis of mercury contaminated samples.

Notes:

General knowledge of analytical chemistry is a significant factor.

Coming soon: a study in the varies approaches of chemical analysis. A "6" folders or parts manual, to be converted to an 800 page reference book by the author.

It must be stated that it is assumed here that the reader is a chemist, or at least an advanced chemistry major university student. It is assumed that the reader is truly familiar with the concepts and the differences between, for example, AAS use, absorption and emission spectroscopy vs ICP use for the same parameter. In fact, solid knowledge of several other areas of analytical chemistry methodology is a must, as Hg analysis will involve a variety of sub-applications.

I was required a decade ago to prepare an analytical laboratory educational or a training manual for a large commercial analytical lab. I had to address, briefly, yet with sufficient illustrations and explanatory notes, all the following issues:

1- Absorption and emission spectroscopy:

- the nature of electromagnetic radiation
- the electromagnetic spectrum
- atomic energy levels
- molecular electronic energy levels
- vibrational energy levels
- raman effect
- nuclear spin behaviour
- electron spin behaiour
- x-ray energy levels

2- Ultraviolet and visible instrumental spectrophotometry:

- radiation sources
- detectors
- readout module
- filters
- monochromators
- grating monochromator system
- instruments for absorption photometry

3- Ultraviolet and visible absorption methods

- fundamental laws of photometry
- spectrophotometric accuracy
- photometric precision
- quantitative methodology
- differential or expanded scale spectroscopy
- difference spectroscopy
- derivative spectroscopy
- photometric titrations
- spectra of solids
- turbidity and nephelometry
- correlation of electronic absorption spectra with molecular structure.

4- Fluorescence and phosphorescence spectrophotometry

- structural factors
- photoluminescence intensity as related to concentration
- instrumentation for phosphorescence measurements

5- Flame emission and atomic absorption spectroscopy
- nebulization
- flame types and temperatures
- interferences
- flame spectrometric techniques

6- Atomic emission spectroscopy

- spectroscopic sources
- atomic emission spectrometers
- photographic detection
- photoelectric detection

7- Inductively coupled plasma ICP

- theory
- comparative study against AAS

8- Infrared spectrophotometry

- correlation of infrared spectra with molecular structure
- instrumentation
- sample handling
- quantitative analysis

9- Raman spectroscopy

- theory
- instrumentation
- sample handling and illustration
- diagnostic structural analysis
- polarization measurements
- quantitative analysis

10- Chemical analysis of surfaces

- ion scattering spectrometry ISS
- secondary ion mass spectrometry SIMS
- auger emission spectroscopy AES
- Electron spectroscopy for chemical analysis ESCA

11- Electron spin resonance spectroscopy

- electron behaviour
- ESR spectrometer
- ESR spectra
- interpretation of ESR spectra
- ENDOR
- ELDOR
- quantitative analysis

12- Nuclear magnetic resonance spectroscopy

- basic principles
- continuous-wave NMR spectrometers
- pulsed fourier transform NMR spectrometer
- spectra and molecular structure
- elucidation of proton NMR spectra
- quantitative analysis

13- X-ray methods

- production of X rays and X-ray spectra
- instrumental units
- radiation measurement detectors
- semiconductor detectors
- direct X-ray methods
- X-ray absorption methods
- X-ray fluorescence methods
- X-ray diffraction

14- Radiochemical methods

- nuclear reactions, radiations and measurements
- applications of radionuclides
- activation analysis

**15- Refractometry and interferometry
polarimetry, circular dichroism, and optical rotatory dispersion.**

- theory
- refractometers
- polarimetry theory
- applications f optical rotatory dispersion and circular dichroism.
- the polarimetter
- instruments for circular dichroism measurement

16- Chromatography

- classification of chromatographic methods
- nature of partition forces
- chromatographic behaviour of solutes
- column efficiency and resolution
- column processes and band broadening
- reduced variables
- time of analysis and resolution
- quantitative analysis

17- Gas Chromatography

- gas chromatographs
- detectors
- optimization of experimental conditions
- gas-solid chromatography

**18- Liquid column chromatography
instrumentation and optimization**

- solvent delivery system
- sample introduction
- separation column

- detectors
- optimization of column performance

19- High-performance liquid chromatography methods

- adsorption chromatography
- liquid-liquid partition chromatography
- ion-exchange HPLC
- exclusion chromatography

20- Mass spectrometry

- components of mass spectrometers
- resolution
- interfacing chromatography and mass spectrometry
- quantitative analysis of mixtures
- use of stable isotopes
- leak detection
- correlation of mass spectra with molecular structure

21- Thermal analysis

- differential thermal analysis
- differential scanning calorimetry
- thermogravimetry
- methodology of DSC (or DTA) and TG
- thermomechanical analysis
- dynamic mechanical analysis
- thermometric titrimetry

22- Electrometric analysis

- electrochemical cells
- electrode potentials

23- pH and ion selective potentiometry

- glass-membrane electrodes
- solid-state sensors
- liquid-membrane electrodes
- gas-sensing and enzyme electrodes
- interferences
- ion-activity evaluation
- problems with pH measurement
- electrometric measurement of pH and pI

24- Potentiometric titrations

- indicator electrodes
- location of the equivalence point
- null-point potentiometry

25- Voltammetry, polarography and related techniques

- characteristics of the dropping mercury electrode
- current-voltage relationship
- the half-wave potential
- amperometric titration methods
- two indicator electrodes
- modern voltammetric techniques
- instrumentation and applications

26- Electrogravimetry and coulometry

Principles, methods and equipment for:
- electroseparations
- electrogravimetry
- electrography

- electrolytic purification
- coulometric methods

27- Electronics use in solid-state instruments

- semiconductor components
- amplifiers
- digital integrated circuits
- digital MSI circuits
- analog MSI circuits
- composite cicuits
- understanding the nature of instrumentation

28- Conductance methods

- conductometric titrations
- electrolytic conductivity
- direct concentration determination
- measurement of electrolytic conductance
- measurement of dielectric constant

29- Process analyzers

- bulk properties
- total peak area vs peak height
- infrared process analyzers
- oxygen analyzers
- on-line potentiometric analyzers
- process gas chromatography
- elemental analyzers

30- Data interprtation

- signal-to-noise ratio
- sensitivity and detection limit
- comparative BDL reporting
- software signal enhancement

- accuracy and instrument calibration
- noise sources
- background absorption
- interferences
- matrix influence at different levels of the parameter
- FAQ and common problems.

This project will consist of a 1000 page reference manual and will be available strictly through Optimum Green Laboratories Canada for an estimated price of $450. A must for analytical laboratories and university libraries.

As I stated, the previous edition of this book, Mercury ultra trace analysis, a limited – special version that was produced for laboratories & reference libraries by Great Canadian Books, was priced at $95 and was sold out in 6 months. This edition is made very affordable for the purpose of wide distribution.

I should also state that the first edition of "As, Se, Sb ultra trace analysis" was sold for $185. It was a limited edition for libraries and laboratories. The Penguin edition later, produced at a larger scale, was made affordable. It is available via www.goudabooks.com

About the author:

Dr. Paul Gouda, C.Chem. P.R.M.D., Ph.D.

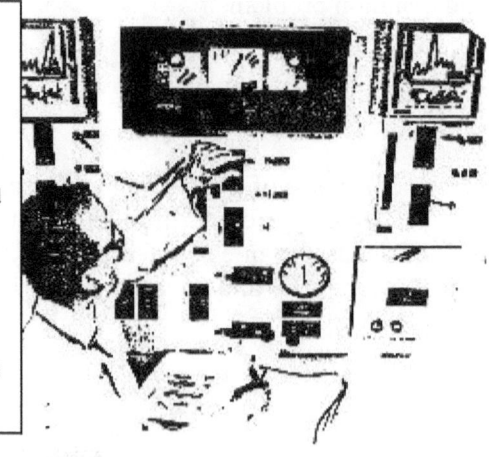

A report by the EPA Gazette featured an interview with Dr. Paul Gouda on ultra trace analysis of food samples and pharmaceutical drugs using a method developed by Dr. Gouda that was named after him, and is now known as {**PGHgICP**).

Dr. Gouda is seen here verifying a result utilizing a state of the art ICP .

*   Chemist Paul Gouda, Ph.D. in analytical, ultra trace analysis.

*   Has written several papers and books on analytical laboratory SOPs, including the analysis of As, Se, Sb, Au, Ag, Hg, Cd, Cu and other elements present in environmental and industrial samples of water, soil, sludge, alloys - at both ultra trace instrumental analysis and bench - wet methods.

* Authored several analytical laboratory procedures including AAS and ICP methods for trace level mercury analysis. As commonly practiced, these methods were officially named after him, and have been globally adopted including by EPL and BRM in Canada.

*   Chief chemist – scientist; Optimum Green Environmental Laboratories.   **www.optimumgreen.com**

www.optimumgreen.com
www.goudabooks.com

gouda@chemist.com

The author is seen here with four research scientists attending his doctorate graduation.

Other books by the author

Poetry:

* Intoxicated emotions
* The scent of the full moon
* The purple dream

Sociology:

* The development of medieval theology, and its reduction to a systematic form.
* The schizophrenic god.
* Survival of the cutest.
* The power of boobs - *"in progress"*
* When the pharaoh created god.
  A second look at the origin of religion - *"in progress"*

Psychology:

* Psychology of officiating, National football "soccer" referee development programme.

Sports & recreation:

* Football (soccer) Referee Instructors & Assessors clinic manual.
* Chef's encyclopedia.

Science:

* Hydrargyrum – Hg analysis from diphenylthiocarbozone to cold vapor.
* Arsenic, Selenium, Antimony - ultra-trace analysis by hydride generation.
* Comparative background interference:
  Atomic absorption VS inductively coupled plasma. Environmental chemical research.
* Pharmaceutical compounds, toxicity.
* Chemical manipulation of hormonal compounds for medical research.
* Choosing the sex of your baby; made simple, made certain. Physiology manipulation in the field of sex selection in mammals – for both, human conception and animal breeding applications.

Search for Gouda Books at:

www.amazon.com

www.bn.com
   "**Barnes & Noble**" USA.

www.chapters.indigo.ca
www.chapters.ca
   "**Chapters – Coles- Indigo**" Canada.

www.biblio.com   (Harvard bookseller)
www.iuniverse.com
www.abe.com
www.alibris.com
www.biggerbooks.com
www.bookdepository.com
www.buy.com - and several other online bookstores.

Notes

www.ingramcontent.com/pod-product-compliance
Lightning Source LLC
Chambersburg PA
CBHW030800180526
45163CB00003B/1102